P9-DXT-444

MATHEMATICAL ASTRONOMY FOR AMATEURS

Books by the same author

A Guide to the Sky
Textbook of Elementary Astronomy
Teaching Astronomy in Schools
The Sky and its Mysteries
Astronomy Old and New
The Young Astronomer
The Solar System

of these exercises, which have been tested and tried in various classes for a wide variety of different types of reader. Optional as the exercises are, however, they do contain information supplementary to the main text, and so each group should at least be read before proceeding to the next chapter, even if the exercises are not actually worked out at the same time.

It is not intended that the absolute beginner should choose this as the *first* book to read; it is meant to be helpful to those who have read one or more of the popular descriptive astronomy books. Nevertheless, there are accounts of familiar topics and these are explained as concisely as possible whenever they are about to be used.

As for precision, angles may be measured to the nearest degree or to a hundredth part of a second; an arithmetical quantity can be worked out to two or ten decimal places. The two kinds of precision are linked: there is no point in calculating to ten figures if the observation on which it was based is known to only two. When a law or a definition is given, can it be regarded as absolutely correct? Perhaps so, within limits; but once it is tested with instruments or mathematics which are of real precision, departures from the law may become evident. In this book we are concerned with principles, and with measurements of the kinds which can be made by amateur astronomers and navigators. Therefore, it has not been thought necessary to give complicated numerical examples, and it may be safely assumed that the laws and definitions are correct. Similarly it has not been thought necessary always to give seconds of time, minutes of arc and so on where a principle can be explained without them, for there is no point in burdening the reader with unnecessary arithmetic.

Metric measurements are used in this book because the International Astronomical Union has requested astronomers to adopt SI units, and they are also being introduced into school science examinations.

Introduction

The aim of this book is to widen the amateur astronomer's understanding of his subject by explaining in the simplest possible terms the 'calculations' behind the most important astronomical laws and principles. Descriptive astronomy, which is well covered by a wide range of publications, excludes to a lesser or greater extent the numerical side of the subject, while textbooks of mathematical astronomy are not designed for the amateur astronomer, unless of course he happens to be a mathematics graduate! This book fills the gap in assuming no more knowledge on the part of the reader than a grasp of 'school' arithmetic and geometry.

A knowledge of numerical astronomy is essential for the reader who aspires to become an active, serious astronomer. Even those astronomers who actively dislike anything to do with mathematics should profit from reading this book, simply because it provides something to do on those frustrating evenings when observation is impossible; observations, after all, have to be interpreted sooner or later. In schools, the book provides examples for all age groups of what can be done outdoors, and of course provides all the necessary numerical information for those taking elementary examinations, such as the British GCE Ordinary Level Astronomy and the astronomy part of the Nuffield Ordinary Level Physics.

The book can be read straight through, some readers progressing faster than others. But there is nothing like working through a number of examples to convince oneself that a particular principle has been understood fully. So for those who wish to consolidate their understanding of the text, exercises and separate answers are given in the appendix. Schools will naturally make greatest use

Contents

First American Edition 1972

ISBN 393 06388 7

Printed in Great Britain

MATHEMATICAL ASTRONOMY FOR AMATEURS

E. A. Beet

W W NORTON & CO INC
NEW YORK

polar axis is 12 714km, whereas at the equator the diameter is 12 756km, but in this book we can regard it as a sphere.

To an astronomer, his position on the Earth's surface is important, and it is defined by two measurements, or co-ordinates, known as *latitude* and *longitude*. The plane of a great circle about a sphere passes through the centre and divides the sphere into two hemispheres; the equator is an example. Fig 1 shows two great circles through the poles, passing through X and G respectively. These are *meridians*. To an observer at X the meridian is his north and south line, though to the astronomer it is not just a line on the ground but a vertical plane reaching to the sky. His meridian provides one of his co-ordinates, but to give it a number there must be a starting point. By international agreement in 1884 the meridian through Greenwich (G) was chosen as the zero, and the longitude of a place is the angle (θ) between the meridian planes through Greenwich and the place concerned. It is measured in degrees up to 180 east and west of Greenwich: thus X is in longitude θ°E. Latitude is the angle at the centre measured north or south along the meridian, from the equator to the place concerned: thus X is in latitude ϕ°N. The line joining places of equal latitude is a parallel of latitude. This is a small circle, the plane of which divides the sphere into unequal parts. North and south latitudes are sometimes written + and –, and so are west and east longitudes. The position of X would then be lat $+\phi°$, long. $-\theta°$.

ANGLES

Everyone is familiar with degrees, 360 to a circle. A degree is divided into 60 minutes of arc, and the minute into 60 seconds. As an example of writing this down, this book is being written in latitude 51°28′53″N. However, angles sometimes have to be expressed in *radian* measure. In Fig 2, AB and CD are arcs of circles having O as centre. The angle θ is defined as the ratio $\dfrac{\text{arc AB}}{\text{radius OB}}$, and in this particular diagram

$$\theta = \frac{20\text{ mm}}{100\text{ mm}} = 0{\cdot}2\text{ radian.}$$

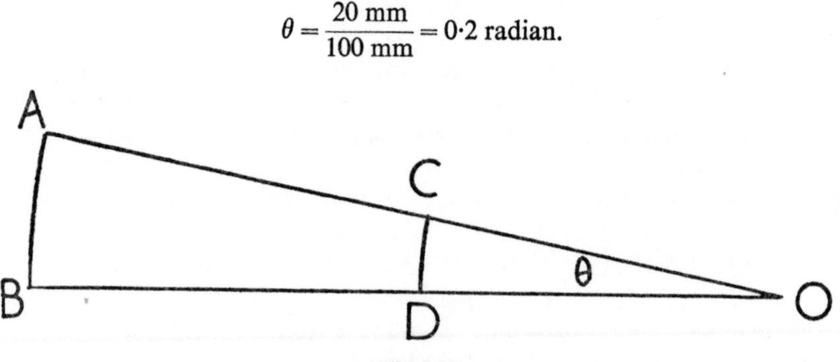

Figure 2

The ratio CD/OD must be the same, of course, as it refers to the same angle. In radian measure, a full circle is 2π, where π is the number 3·1416 (approximately $\frac{22}{7}$), used for calculating circumferences and areas. Thus

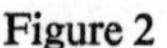

$$360^\circ = 2\pi \text{ radians}; \quad 1^\circ = \frac{2\pi}{360} \text{ and } x^\circ = \frac{\pi x}{180} \text{ radian.}$$

(It follows, too, that 60′ and 60 × 60″ also equal $\pi x/180$.) In reverse: 2π radians = 360°; x radians = $180x/\pi$ degrees.

Referring again to Fig 2, we now know that AB/OB = CD/OD, where AB and CD are arcs. When an astronomer is observing the angular size of a celestial object it is usually small and the arcs are therefore indistinguishable from straight lines. Thus AB would be the linear diameter of the object and OB its distance away, and by the rules for similar triangles we again have

$$\frac{AB}{OB} = \frac{CD}{OD} \quad \text{and also} \quad \frac{AB}{CD} = \frac{OB}{OD}.$$

Say, for example, that you find that a disk 2cm in diameter just covers the Moon when placed 230cm from the eye. The value for the angular diameter of the Moon would be:

$$\text{angle} = \frac{\text{arc}}{\text{radius}} = \frac{2}{230} = 0{\cdot}0087,$$

and this becomes

$$0{\cdot}0087 \times \frac{180 \times 60}{3{\cdot}1416} = 29'{\cdot}9.$$

It is not really necessary to work out the 0·0087 separately.

What would be the linear diameter of the Moon if its distance is taken to be 370 000km? Using the rules of similar triangles:

$$\frac{\text{diameter}}{\text{distance}} \text{ of Moon in km} = \frac{\text{diameter}}{\text{distance}} \text{ of disk in cm},$$

$$\text{diameter of Moon} = \frac{2}{230} \times 370\,000 = 3\,218\text{km}.$$

Before leaving the matter of angles, it must be pointed out that when adding or subtracting them it is sometimes necessary to subtract or add 360. For instance:

$$147° + 202° = 349° \quad \text{but } 147° + 232° \ (-360°) = 19°.$$

Similarly, as we shall see later, when adding or subtracting times it may be necessary to subtract or add 24 hours.

THE MOTION OF THE EARTH

There is nothing at rest in the universe, so whenever we talk about motion there must be a stated or inferred frame of reference. At this moment we are probably inferring that the floor is at rest, and any motion around the room is 'relative to the floor'. Taking the Earth as the fixed point, the Sun appears to move around it once a year, passing through the twelve constellations of the Zodiac along a line called the *ecliptic*. If we take the Sun as fixed, then the Earth is in orbit around it; the plane containing the orbit—the paper on which it might be drawn—is the *plane of the ecliptic*. These two modes of thought are called *geocentric* (Earth-centred) and *heliocentric* (Sun-centred), terms which will turn up again.

The Earth's axis of rotation is roughly at right angles to this plane, but differs from perpendicularity by an angle of $23\frac{1}{2}°$; this quantity is the *inclination of the axis*, and is also known as the *obliquity of the ecliptic*, ϵ (23°26′35″, 1970). A spinning top remains upright as long as the spin is fast enough, and if it is a well-made gyroscope top it will maintain the direction of the axis of spin pretty well even when not quite upright. The Earth also maintains its axis of spin; the axis produced above the north pole points to the *celestial pole*, near which is the familiar Pole Star. When, however, the axis of spin of the gyroscope top is not vertical, or if the balance of a vertical one has been deliberately disturbed by adding a very small weight to one side of the frame, an additional phenomenon appears. The axis will show what might be called 'a systematic wobble', its end slowly tracing out a circle in space. This phenomenon is known as *precession*, and the Earth has it too. The axis is inclined at $23\frac{1}{2}°$ to the perpendicular to the ecliptic (Fig 3), but its direction relative to the stars moves around this perpendicular once in 26 000 years—too slowly to worry us at present.

The apparent size of the Sun is not constant, so the Sun's distance from the Earth is not constant either; it varies from 147×10^6 to 152×10^6km, the mean distance being 149·6 million km or approximately 93 million miles. The points in the orbit at

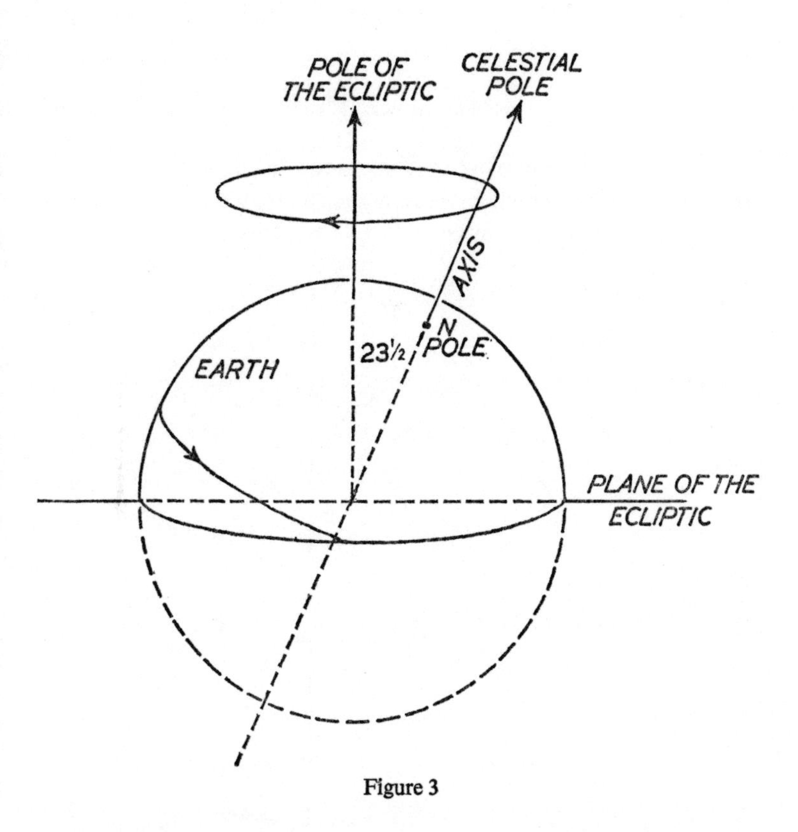

Figure 3

which the Earth is nearest and farthest from the Sun are *perihelion* and *aphelion* respectively, collectively called the *apses*. The orbit, therefore, is not a circle with the Sun at the centre. It is very nearly a circle with the Sun off-centre (eccentric), and is actually an ellipse with the Sun at one focus. As the ellipse plays a very important part in astronomy, it is important to understand thoroughly its properties and construction.

THE ELLIPSE

The ellipse is a symmetrical oval, as shown in Fig 4. In the case of the Earth's orbit, which is much nearer a circle than as drawn in this figure, the Sun would be at *F*. The perihelion is at *A*, and the

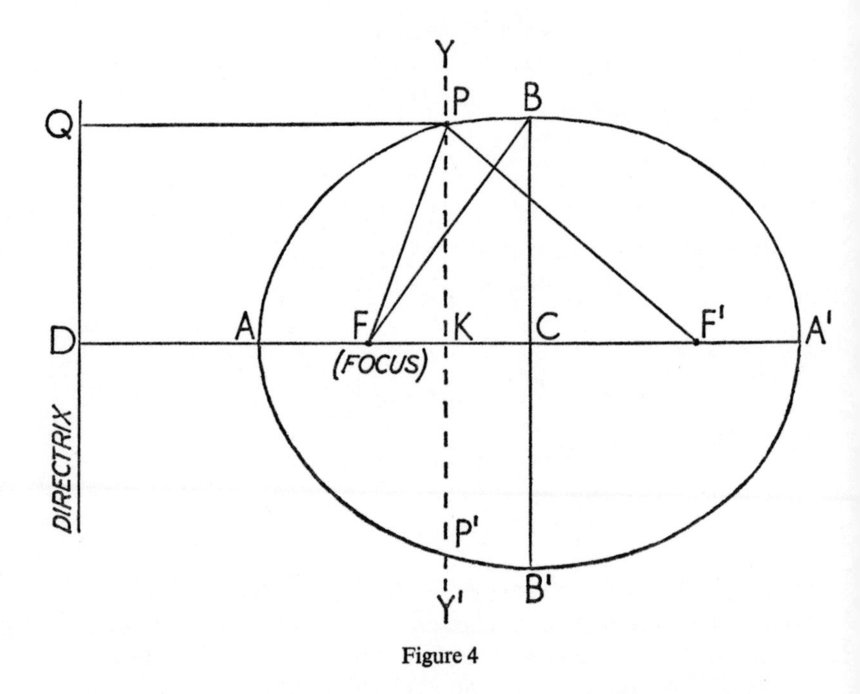

Figure 4

perihelion distance FA is in astronomical tables represented by *q*. *AA'* is the major axis, and the *semi-major axis* CA is *a* in the tables and is also the familiar mean distance of 93 million miles. *BB'* is the minor axis and CB is represented by *b*. An ellipse is a curve such that for any point on it (P in Fig 4) the ratio of its distance from a fixed point, called the focus, to its perpendicular distance from a fixed straight line, called the directrix, is a constant with a value less than 1. This is a definition; the additional facts which follow are derived from it in mathematics books. The constant is the *eccentricity* e. (In a circle, $e = 0$.) Thus,

$$\frac{PF}{PQ} = e, \quad \text{and similarly} \quad \frac{AF}{AD} = e.$$

There is a second focus F′, and the distance of either focus from the centre C, CF or CF′, is equal to ae. The sum of the distances from any point to the two foci is a constant. For the point A, its distance from F is $a - ae$ and from F′ $a + ae$, so the sum is $2a$. Thus $FP + PF' = 2a$ and as B is equidistant from both foci, $FB = a$.

DRAWING AN ELLIPSE

There are three main methods in drawing an ellipse, given a and e.

Method 1

Draw the major axis (Fig 4) and insert F, the Sun in the case of a planet orbit. AF is $a - ae$; insert A (if $a = 10$cm and $e = 0{\cdot}6$,

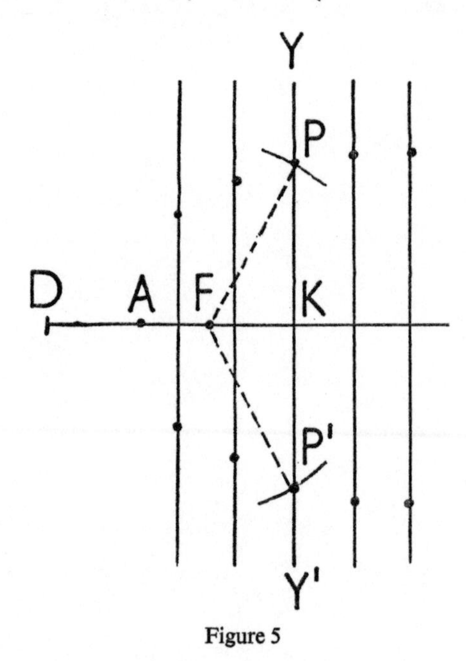

Figure 5

$AF = 10 - 10 \times 0{\cdot}6 = 4$cm). DA is AF $\div\, e$; insert D ($DA = 4 \div 0{\cdot}6 = 6{\cdot}7$cm). Draw a number of lines (Fig 5) perpendicular to the axis, of which YY′ is one. With centre F and radius DK $\times\, e$ (DK is equal to the perpendicular distance PQ) draw arcs to cut your perpendicular in P and P′. These are two points on the curve. Repeat for other lines until you have enough points to draw the ellipse. This method is recommended when only a part of the orbit is wanted, such as that of a comet when near the Sun, but is inconvenient for getting a complete orbit.

Method 2

Place your paper on a drawing board or a flat sheet of thick cardboard. Draw the major axis. (Fig 4.) Insert F. FF′ $= 2ae$ (in the previous example, $2 \times 10 \times 0{\cdot}6 = 12$cm); insert F′. Through F and F′ stick pins into the board. Prepare a loop of cotton or thin string of length, after the knot has been tied, equal to $2a + 2ae$. Place the loop over the two pins, draw it tight with the point of your pencil, and then move the pencil around the pins keeping the loop tight all the time.

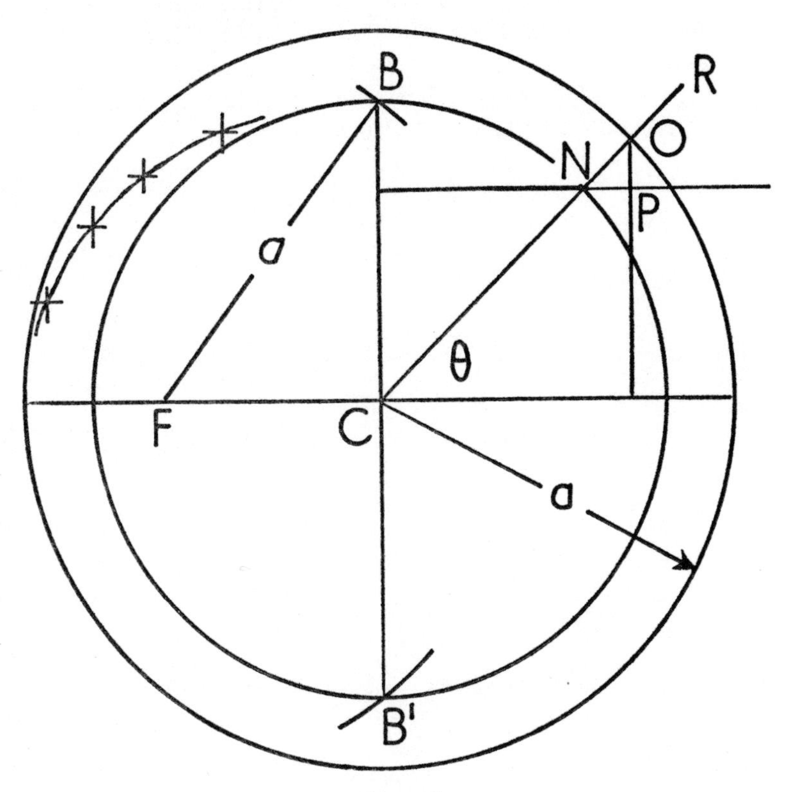

Figure 6

Method 3

An alternative to Method 2 that is more trouble but also more accurate (see Fig 6). Draw the major axis and insert F. FC = ae; insert C and through it draw a perpendicular to the axis. With centre F and radius a draw arcs to cut the perpendicular at B and B′; these are the ends of the minor axis. With centre C draw circles of radii a and CB. Draw any radius CR, cutting the circles at N and O. Through N draw a line parallel with the major axis, and through O one parallel with the minor axis; their intersection P is a point on the ellipse. Repeat until enough points have been obtained to draw the curve. When drawing a line through a series of points, always keep the hand on the inside of the curve and move the paper as may be necessary. The reader should be able to see for himself what properties of the ellipse have been used in methods 1 and 2, but to justify method 3 requires more mathematics than is assumed in this book (see Note 3, page 104. Notes 1 and 2 give yet another method of drawing an ellipse.)

THE EARTH'S ORBIT FROM FIRST PRINCIPLES

The ellipses just considered have been drawn from data already known, the size being determined by a and the shape by e. An alternative way, when these quantities are not known, is to work from observational material. Measurement along the line of the ecliptic is called celestial longitude, and is made eastward up to 360° from a starting point known as the *First Point of Aries*, denoted by the sign ♈. How that point is chosen will be explained when we come to the celestial sphere in Chapter 3. In the table below, column A gives the position of the Sun as seen from the

TABLE 1

Date		A	B	Date		A	B
1970	Jan 1	280°	32′·6	1970	July 1	099°	31′·5
	Feb 1	312	32·5		Aug 1	128	31·6
	Mar 1	340	32·3		Sep 1	158	31·7
	Apr 1	011	32·0		Oct 1	187	32·0
	May 1	040	31·8		Nov 1	218	32·3
	June 1	070	31·6		Dec 1	248	32·5

Earth, *geocentric longitude*. The direction of the Earth relative to the Sun, *heliocentric longitude*, will be this ±180° because the imaginary observer on the Sun is viewing the opposite way along the same line. Column B gives the corresponding angular diameters of the Sun as seen from the Earth. From Fig 2 it should be clear that as the diameter represented by AB is constant, the angle must be inversely proportional to the distance AO, and, what comes to the same thing, the distance is directly proportional to the reciprocal of the diameter. The information has been compiled from the *Astronomical Ephemeris*, as it would be difficult, but not impossible, to get it from personal observation. (Tricker in

The Paths of the Planets deals with this kind of problem very fully, including the observational side.)

Take a large sheet of paper, put the Sun S near the middle (Fig 7) and rule SZ as a zero line. As an example of procedure, take March 1. The heliocentric position of the Earth will be $340 - 180 = 160°$. Measure this angle anticlockwise, ZSX. Find

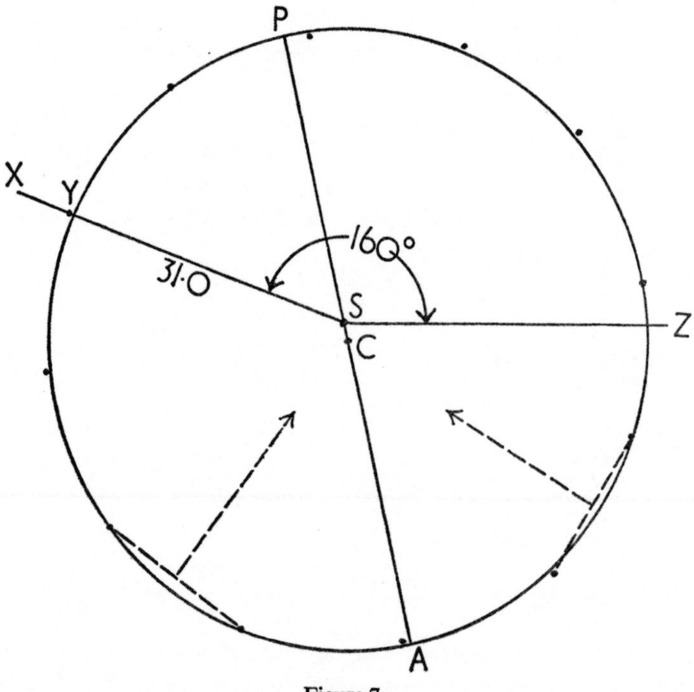

Figure 7

the reciprocal of 32·3 from a table book, a slide rule or by common arithmetic; 1/32·3 comes to 0·031. Now choose a multiplying factor suitable for your paper. If we choose 1 000 the reciprocal becomes 31·0, and this expressed in cm would be a reasonable size. It must not be less than this; use a larger multiplier if the paper will take it. Make SY = 31·0cm. Then Y is a point on the orbit. Proceed similarly with the other eleven points. Then *either:* keeping your hand on the inside draw a smooth curve through the points. By trial, locate and draw the longest diameter through S. This is the major axis PA; mark the mid-point C. *Or:* The orbit is very nearly a circle, so use a large compass, or improvise with a pin and a piece of string, and draw the circle which most nearly fits the points. The centre and radius must be found by trial, but an approximate centre to start from can be found from the intersection of the perpendicular bisectors of two chords (broken lines in Fig 7). The line CS produced can be taken as the major axis. Note that Fig 7 is illustrative only; it is on too small a scale to have been drawn from the above table.

Having drawn the orbit, measure the angle ZSP, where P is the end of the major axis nearer to S; this is the *longitude of the perihelion*, ϖ (another form of pi), in astronomical tables. Measure a (half the major axis) and CS. From CS $= ae$ calculate e. When you have done so, and not until, check from the table on page 61. We have now found the shape and orientation of the Earth's orbit, but not its real size, as the multiplying factor was an arbitrary one, not a scale of kilometres.

THE MOON

The Earth is accompanied by a satellite, the Moon, which is a sphere of diameter 3 476km situated at a mean distance of 384 000km. As this distance is very small compared with the radius of the Earth's orbit, the path of the Moon relative to the Sun is very much the same as that of the Earth and, like it, is always concave to the Sun. The two orbits are interlaced; at new moon that of the Moon is nearer the Sun, and at full moon it is the further one. The ellipse which we regard as the orbit of the Earth is really that of the combined centre of gravity of the two bodies, the *barycentre*, but remembering the remarks in the introduction about precision we will not pursue this topic further.

Relative to the Earth, the Moon is moving in an ellipse with the Earth at one focus; the nearest and farthest positions are called *perigee* and *apogee*. As with the Sun, its angular diameter varies with distance, and its geocentric longitude relative to the stars can be measured. Thus the orbit can be drawn in the same way as the last exercise, and data for doing so will be found in Exercise 16. The orbital eccentricity is larger than that of the Earth, so you will have a better chance of getting it right.

The time taken to go around the Earth, measured against the background of stars, is the *sidereal period*, which is 27·32 days. The Moon also rotates on its axis in 27·32 days, so by the time it has made half a rotation it is on the opposite side of the Earth and we see the same hemisphere of our satellite as we did before—the only side we ever see without a space ship. A phenomenon known as libration does enable us to see a little more than one hemisphere, but this will be left to your other reading. Your other reading should by now have explained the phases of the Moon, so that you will realise that 'new moon' occurs when the longitude of the Moon is the same as that of the Sun—as seen from the Earth they are in the same direction, though not necessarily in the same straight line. The next new moon is not 27·32 days later; the sidereal period is not what we ordinarily call a month. The Moon moves around the ecliptic at an average speed of 360/27·32 =

13·18 degrees per day. At the same time the Sun is travelling at 360/365 = 0·986 degrees per day. Thus the Moon is gaining on the Sun by 12°·19 per day. From one new moon to the next it must gain 360°, and this would take 360/12·19 = 29·54 days. This is the *synodic period*, or a lunar month. A further thought about the 12° eastward motion with respect to the Sun is that the Moon will cross the observer's meridian later each day by the time it

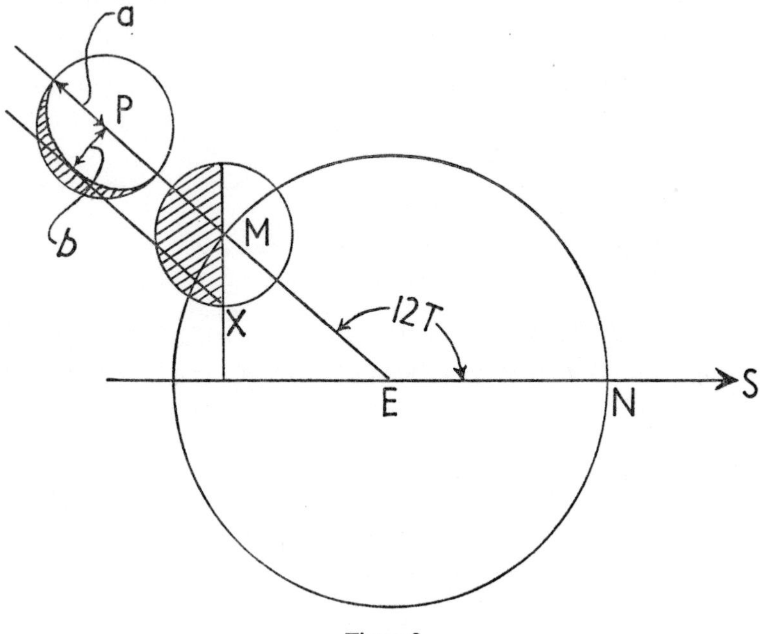

Figure 8

takes the Earth to rotate the extra 12°. This will be $24/360 \times 12 = 0{\cdot}8$ hour or 48 minutes. Hence the very general statement that moonrise and high tide get 50 minutes later each day—very general because both moonrise and high tide are affected by other factors as well this average 12°.

Suppose you wish to know what the phase of the Moon will be on some future date. A friend is coming on Friday and hopes to see the Moon through your telescope. 'What shall I be able to show him?' 'Where will the terminator (boundary between light and darkness) be on that evening?' Proceed as follows, Fig 8. E is the Earth, ES is the direction of the Sun, and the orbit of the Moon is drawn as a circle. Let T be the time in days which has elapsed since new moon (position N). Then the Moon will have moved by 12T°, so make the angle NEM = 12T°. Draw a circle about M and another on EM produced. Through M draw a perpendicular to ES, cutting circle M at X. Then X is the boundary of the illuminated hemisphere. Through X draw a line parallel with EM, and sketch in the terminator as shown. The terminator is really an ellipse, which is difficult to draw on a small scale, so if you want an accurate one measure *a* and *b* and draw the ellipse on a larger scale by method 3, page 12. A final word about this daily 12°. It is an average value, for the motion of the Moon is not uniform, so small discrepancies will occur between your predictions and what you observe.

The orbit of the Moon does not lie in the plane of the ecliptic, but is in a plane of its own inclined at about 5° with it. The two planes intersect in a straight line through the Earth meeting the orbit in two points called nodes. Where the Moon travels from south to north of the ecliptic plane is the *ascending node*, represented by the symbol ☊; where it moves south again is the *descending node* ☋. Just to complicate the issue, the node line is not fixed in space, but rotates in a westerly direction in a period of 18·6 years. Eastward motion in our sky, or anti-clockwise rotation when viewed from a very distant space ship over the north pole, is said to be *direct*, and the reverse is *retrograde*. In our present Earth-centred frame of reference the Sun has a direct motion of 360° per year, and the nodes a retrograde one of 360/18·6 = 19°·4 per year. Thus they separate at a rate of 379°·4 per year; relative to the nodes the Sun moves 379·4/365 × 29·5 = 30°·6 per synodic month from new moon to new moon or 15°·3 from new to full. These facts have an important bearing on eclipses.

At new moon the longitudes of the Sun and Moon are equal (conjunction) and at full moon they differ by 180° (opposition), but there will not normally be an eclipse at either, because of the inclination of the Moon's orbit. Theoretically the events would have to occur exactly on the line of the nodes for an eclipse to happen, but as the Sun and Moon are not points but have a finite size, about ½° each, there is some latitude—the new or full moon need be only near the node. Solar eclipses can occur within about 16° of a node, and lunar within about 10°. These are rough average values to work with; the actual values are outside the 'precision limit' of this book (see Note 4, page 105). If new moon occurs when the Sun is within the 32° arc in Fig 9 (16° on either side of the node) there will be a solar eclipse, position A. From one new moon to the next, 29½ days later, the Sun moves on by less than 32°, so if there is no eclipse in one lunation there must be one in

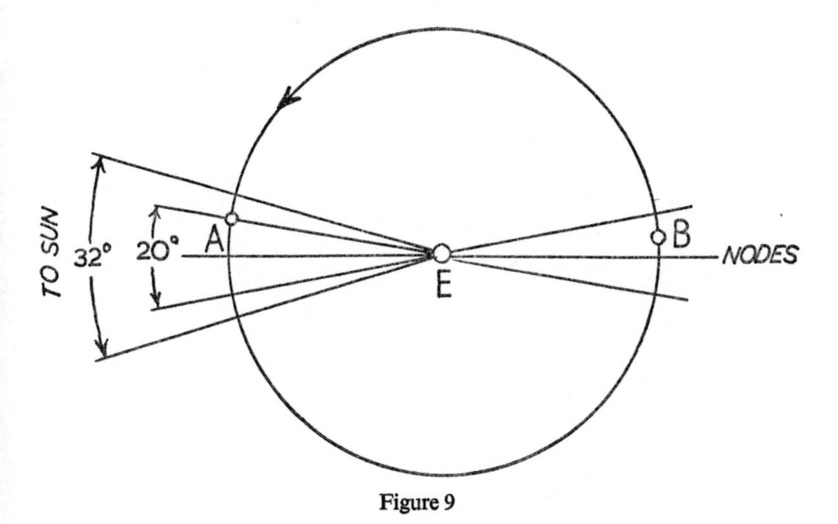

Figure 9

the next. If there is an eclipse at A, then while the Moon is going round to its full position the Sun moves on 15°·3; it is still within the 20° arc, so a lunar eclipse occurs at B. On the other hand, if A were a few degrees further on, B would fall more than 10° from the node and there would be no second eclipse. If A were slightly earlier than shown, just within the solar limit, the Sun's position 30°·6 on at the next new moon would still be within the limit, and another solar eclipse would occur. Thus there must be one eclipse and there might be two or three. Six lunations later the Sun will have moved $6 \times 30{\cdot}6 = 183{\cdot}6°$ and is in the eclipse zone at the other end of the node line; there is another eclipse 'season'. So, then, minimum two (both solar), maximum six. Twelve lunations = $12 \times 29\frac{1}{2} = 354$ days, and as this is just less than a year it is possible for one eclipse of a third set to fall within a year, absolute maximum seven. The limiting cases do not happen very often; there were seven in 1935 and two in 1969. Penumbral eclipses of the Moon have not been considered in this discussion, and are hardly noticed when they do occur; and it is assumed that distinction between total and partial eclipses is known from previous reading. There are other interesting things about the arithmetic of eclipses, and readers wishing to go deeper into the subject should refer to more detailed works.

2
Time

SIDEREAL, SOLAR AND MEAN TIME

TIME-ANGLE CONVERSION

LONGITUDE AND TIME

SIDEREAL-MEAN TIME CONVERSION

THE YEAR

Timing an object when it is crossing the meridian is called observing a transit, and is done in observatories with an instrument called a transit circle, shown diagrammatically in Fig 10. The telescope is mounted on an east–west axis between two pillars so that its motion is confined to the north–south vertical plane,

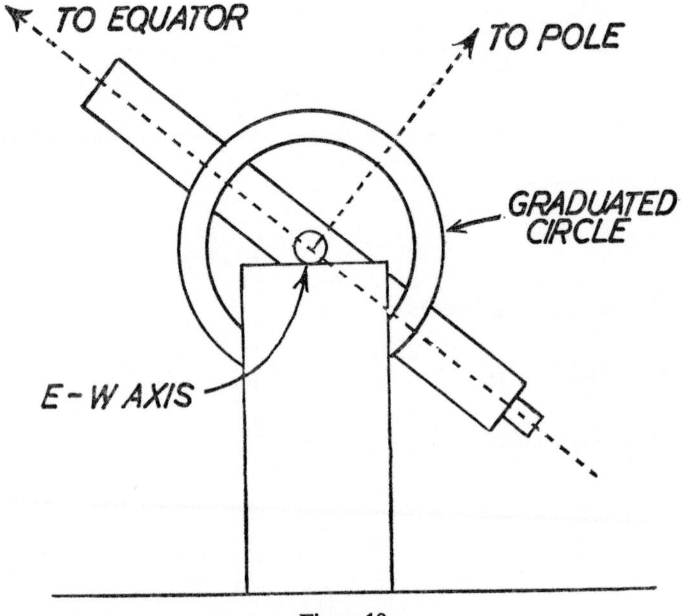

Figure 10

that is, to the meridian. There are crosswires in the eyepiece, and when a star is seen on the central vertical wire it is on the meridian, no matter what the altitude of the star may be. The telescope carries a vertical graduated circle so that its position in the vertical plane can be recorded (Fig 10). It reads zero when the telescope is at 90° with the celestial pole; all stars passing across its axis must also be 90° from the pole—they lie on the *celestial equator*. As with much other technical equipment, the operation of a modern observatory transit instrument is now largely automatic, thereby eliminating personal errors on the part of the observer.

SIDEREAL, SOLAR AND MEAN TIME

The rotation of the Earth upon its axis has been the basis of time measurement for very many centuries, the 'day' being the interval between two successive transits of the same object. If that object is a star, the interval is the *sidereal day*, divided into hours, minutes and seconds in the usual way. The sidereal clock in the observatory shows these hours, but as with longitude we must specify a zero point. For this purpose the zero is the same as for celestial longitude (page 13), the First Point of Aries. When ♈ is crossing the meridian the sidereal clock reads 0h. When the clock reads 3h, ♈ will have moved westward by $3 \times 15 = 45°$ and is said to have an *hour angle* of 45° or of 3h, whichever expression happens to be the more convenient. Thus *sidereal time*, as indicated by a sidereal clock, is defined as the hour angle of the First Point of Aries (H.A.♈). The actual measurement of hour angle, in degrees along the celestial equator, does not concern us in this chapter—we shall look at the clock.

Suppose that we are observing the transit of the Sun. A transit is observed; one sidereal day later the transit telescope will again point in the same direction in the heavens, but not at the Sun, for that body has in the interval moved just about a degree to the east of its original position. The second transit of the Sun will not occur until the Earth has rotated another degree, which takes 4 minutes. Thus the *solar day* is approximately 4 minutes longer than the sidereal day. When the Sun is on the meridian the solar time is traditionally 12h, not zero, so *apparent solar time* as recorded by a sundial is the hour angle of the true Sun (HATS) + 12h. It is called 'apparent' because it is what appears on the sundial; apparent noon was the moment of the shortest shadow in Ex. 1, but this is not the solar time by which we normally reckon. The Sun's daily motion along the ecliptic is not uniform; the 0·986 of a degree in the last chapter and the 'another degree' in this paragraph are average values. Neither is the Sun normally on the celestial equator, whereas ♈ is. Thus the solar days are not uniform in length, and are incompatible with mechanical clocks and the terribly time-conscious era in which we live. The true Sun has been replaced by a fictitious body called the mean Sun which moves along the equator at a uniform daily rate, making all days the same length. Thus our clocks keep *mean time*, which can be defined as the hour angle of the mean Sun (HAMS) + 12h. As everyone knows, local times in different places do not agree, on which more later. For announcing or recording astronomical events a standard time must be specified, and that of Greenwich, *Greenwich Mean Time* (GMT), has been adopted for astronomers and navigators the world over, though astronomers now call it *Universal Time* (UT) (see Note 5, page 105). In this book we shall sometimes use one name and sometimes the other so that the reader becomes thoroughly familiar with both. Mean time is, of course, based on the rotation of the Earth, and the steadily improving time-keeping devices in the laboratories have revealed irregularities even in that. Mean time is not absolutely uniform, so computers (meaning human ones, though it applies also to the electronic variety) work with a perfectly uniform system called *Ephemeris Time*. The difference is small, and ordinary astronomical work and navigation is done in GMT (UT); it is mentioned here only in case the reader comes across the expression elsewhere and wonders what it means.

The actual observer cannot look at the fictitious mean Sun; he must use the real one, so he will need to know the difference between apparent and mean solar time. This difference is known as the *equation of time*. There is some inconsistency over the use of + and –; in this book we shall call it + when the sundial is fast compared with the clock. Thus it is the quantity to be added to mean time to get apparent time:

$$\text{mean time} + E = \text{apparent time}$$

or

$$E = \text{HATS} - \text{HAMS}.$$

If the sundial shows 11 am when local mean time is 10-55, then we must add 5 minutes to the clock, and the equation is +5m. To take an example, when is the Sun true south if the equation of time is –7m? This means that the sundial is 7m slow and the clock will read 12-07 by the time the Sun is on the meridian; or

$$-7m = 12h\ 00m - HAMS,$$
$$HAMS = 12h\ 00m + 7m = 12h\ 07m.$$

When extracting the equation of time from an almanac, first find the note—there is sure to be one—stating how the compilers have applied the sign. It is becoming more usual to tabulate the mean time of the transit of the true Sun (= 12h apparent solar time) for the meridian of Greenwich, which is probably what you really want to know. The equation of time is zero four times a year; on these four dates local mean time is the same as sundial time. (For explanation see one of the more advanced books, such as Barlow and Bryan.) The general behaviour of E is given in Table 2.

TABLE 2

Approx. date	Sign of E	Sundial	Difference
Feb 11	−	slow	14m
Apr 15	0	correct	0
May 14	+	fast	4m
June 14	0	correct	0
July 26	−	slow	6m
Sep 1	0	correct	0
Nov 3	+	fast	16m
Dec 25	0	correct	0

TIME-ANGLE CONVERSION

In one sidereal day the Earth rotates 360° with respect to a star, so in one sidereal hour the rotation is $360 \div 24 = 15°$. In one mean time day, it is true that with respect to the stars the Earth has rotated by more than 360° and has taken longer in doing it. Nevertheless the rotation is still 360° with respect to the mean sun, and that will be 15° per mean time hour. Some people are not satisfied that the 15° per hour applies to either kind of time, but it does. It is frequently necessary to convert angle into time and vice versa.

Example 1: Convert 5h 14m 29s into arc.

15° per hour	$5 \times 15 =$	75°
$\frac{1}{4}$° per min.	$\frac{1}{4} \times 14 =$	3° 30′
$\frac{1}{4}$′ per sec.	$\frac{1}{4} \times 29 = 7\frac{1}{4} =$	7′ 15″
		78° 37′ 15″

Example 2: Convert 18°20′16″ into time.

4 min. per 1°	$4 \times 18 = 72 =$	1h 12m
4 sec. per 1′	$4 \times 20 = 80 =$	1m 20s
$\frac{1}{15}$ sec. per 1″	$\frac{1}{15} \times 16 =$	1·06
		1h 13m 21s

An alternative method is

$$\text{h m s of time} \begin{Bmatrix} \times 15 \rightarrow \\ \leftarrow \div 15 \end{Bmatrix} \text{deg. min. sec. of arc}$$

but you are less likely to make mistakes if you set out the work in full as in the examples.

LONGITUDE AND TIME

Everyone knows that time is not the same everywhere. In Fig 11 the circle is the Earth's equator, the centre is the N pole, and the radial lines are meridians. The Sun is presumed to be on the right

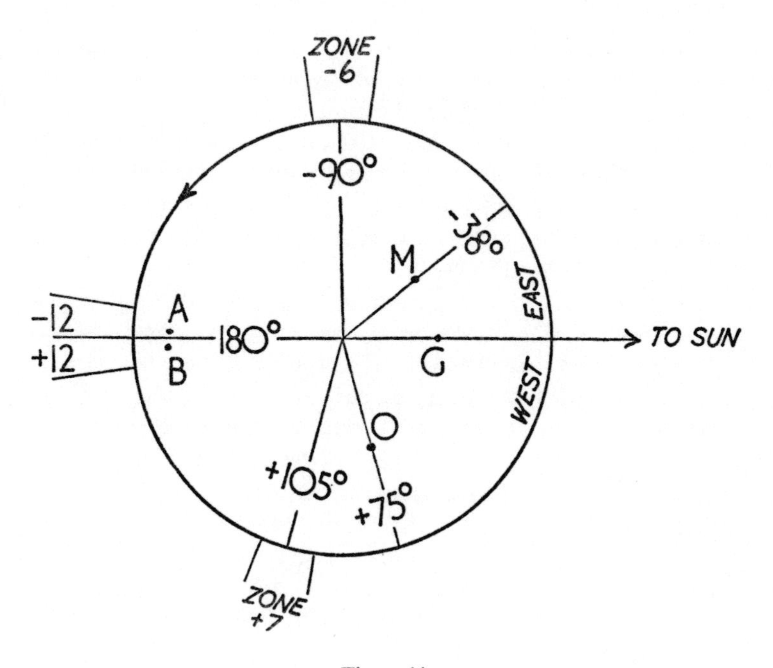

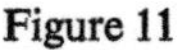

Figure 11

of the diagram, so at Greenwich, G, the time is 12h. At Ottawa, O, it is certainly not, and will not be until the Earth has rotated by the longitude of 75° (actually 75°43′). This will take 5h, so the time at Ottawa is $12 - 5 = 07$h. In the case of Moscow, M, long. 37°34′E, it is afternoon; their meridian passed the Sun some time ago. Converting the longitude to time we get 2h 30m 16s, so when it is noon at Greenwich the *local mean time* (LMT) in Moscow is 2h 30m 16s. In general

$$\text{LMT} = \text{GMT}\begin{cases} + \text{ long. E} \\ - \text{ long. W} \end{cases}$$

when longitudes are expressed in time. If we use the conventional signs + for west and – for east we get the even more general equation

$$\text{GMT} = \text{LMT} + \text{long.}$$

This seems to be the moment to let out a bee that has been buzzing in the writer's bonnet for some forty years as a physics teacher. Don't become a slave to the god Formula. So often the immediate reaction to any problem has been: 'Please, sir, what's the formula?' *Think* what you are doing; mechanically substituting in a formula may get an answer, but will not get understanding. In time conversions, always begin with the question: 'Is the time I want earlier or later than the one I've got?'

For administrative convenience a whole area keeps standard time regardless of the LMT of the various places. Great Britain, for instance, nominally keeps GMT throughout. Very large areas, such as the oceans, the USA and the USSR are divided into 15° strips each keeping the time of its central meridian. For example (Fig 11) longitude 105°W is 7h behind Greenwich, and in Zone +7, from 97½° to 112½°, all clocks should read 7h slow on GMT. If an amateur astronomer sees a meteor at 01h 43m, he knows the event to be at 01h 43m + 7 = 08h 43m UT. Similarly a resident near long 90° in Zone –6 must subtract 6 from the LMT of his zone to get UT. If he made his observation at 01h on Aug 1 his UT would be 19h on July 31. In general

$$\text{GMT (UT)} = \text{zone mean time} + \text{zone number.}$$

In countries of ordinary size it is usual to use national boundaries rather than the zone meridians to mark their standard time, and in some cases the time kept is not that of the zone concerned. Britain is in Zone 0, but for many years we have kept Greenwich + 1 in the summer months (*summer time*). Astronomers do not use it, but they must remember that in summer their watches are an hour fast.

Let us now consider meridian 180° (Fig 11 again). A is just inside Zone −12, and B in Zone +12. If it is 12h on Tuesday the 8th at Greenwich, for A it is, say, 23h 59½m on Tuesday and Wednesday is about to begin. It happens to be his birthday, so he can hop over the meridian to B and have Tuesday all over again, for the time there is 0h 0½m. On the other hand B, who has a Tuesday appointment with his dentist, can hop over the line and in a few seconds it will be Wednesday the 9th. Meridian 180° is the *International Date Line*, but with variations to avoid inhabited areas (and trickery with birthdays and dentists). When ships cross the line from west to east, like A, the calendar is put back one day, and when crossing from east to west, like B, it is put on one day.

SIDEREAL-MEAN TIME CONVERSION

Our daily routine is governed by mean time and our study of the night sky by sidereal; it is therefore necessary to be able to change from one to the other. For choosing a star map, and ordinary naked eye or binocular observation, we need it only to the nearest half hour or so, and it can be estimated without the use of an almanac. When the Sun in its annual journey around the ecliptic passes ♈ in March, both will be on the meridian about the same time; the sidereal time will be 0h but the solar time 12h. When the longitude of the Sun is 180° on September 23 the sidereal time of its transit will be 12h, so sometime on that day (for an estimate like this it does not matter when) the mean and sidereal clocks will agree. We have seen that the sidereal clock gains 4m a day, which is just about 2h a month, so we can allow 2h for each complete month and 4m for each extra day, and ignore the different lengths of the two kinds of hour.

Example: estimate the sidereal time at 19h 30m on Feb 2

Sep 23 at 0h	ST =	0h	0m
Jan 23 at 0h (+4 months)		8	0
Feb 2 at 0h (+10 days)			40
19h 30m add (and −24)		19	30
		4h	10m

For more accurate work, such as setting the circles of an equatorial telescope, the starting point is the sidereal time at Greenwich at the previous midnight (0h UT), as given in an almanac. The sidereal day is 23h 56m 4s mean time, so a sidereal clock gains 3m 56s per day, or in one hour 236 ÷ 24 = 9·8s. For the present purpose we can call it 10s per hour and neglect parts of an hour. (If you wish you can add another second for each 6m, though this is an over-correction. Almanacs usually include tables for applying these corrections.)

Example 1: Find the sidereal time on Feb 2 at 19h 30m GMT.

Feb 2 at 0h UT, from almanac		ST = 8h 47m 5s
Mean time interval	19h 30m	
add 10s per hour	3m 10s	
Sidereal interval	19h 33m 10s	19h 33m 10s
add (and −24)		4h 20m 15s

Compare with the estimate made for the same date and time.

Example 2: At what GMT on Feb 2 will it be 3h 15m 10s ST?

Sidereal time required	3h 15m 10s
Sidereal time at 0h UT	8h 47m 5s
Sidereal interval: subtract (and +24)	18h 28m 5s
subtract 10s per hour	3m 0s
Mean time interval since 0h which is the required mean time	18h 25m 5s

Suppose that you were not on the Greenwich meridian, but in long. 3°W? At 19h 30m GMT your LMT will be 19h 30m less 4 × 3m = 19h 18m, and this becomes the mean time interval (in Example 1) from local midnight to the time of observation. If, like this, you are not far from the Greenwich meridian, the sidereal time will not change much between Greenwich midnight and yours, but for more distance places proceed as follows.

Example 3: Find the sidereal time at 19h 30m LMT in long. 77°E.

77° converted to time = 5h 8m
GMT is earlier by 5h 8m = 14h 22m

By the method of Example 1 (check it yourself)

Greenwich sidereal time	= 23h 11m 25s
add long. E (and −24)	5h 8m 0s
Local sidereal time	= 4h 19m 25s

Notice that the error which would have occurred by using Example 1, with the local mean time as the interval, is less than a minute—but then 77° is less than half-way around the world.

THE YEAR

We have already noted that the celestial equator is everywhere 90° from the poles, and that ♈ lies on the equator. When the Sun passes that point it too is on the equator, and this mid position between the poles makes day and night equal in length. The event is called an *equinox*, and this one in March heralds the beginning of spring; there is another in September marking the first day of autumn. There are four named 'markers' along the ecliptic, the dates being approximate:

Longitude	0°	spring equinox	Mar 20
	90°	summer solstice	June 21
	180°	autumn equinox	Sep 23
	270°	winter solstice	Dec 22

The *tropical year* is the interval between one spring equinox and the next; it is 365·242 days, or 365d 5h 48m 46s. This is longer than the *civil year* of 365 whole days, so the exact time of an equinox will get later year by year. Here for instance is a series for the autumn event:

1960 September	23d 01h 00m
1961	06h 43m
1962	12h 35m
1963	18h 24m
1964	00h 17m

At first sight it looks as though it had moved on to September 24 by 1964, and if nothing was done about it the first day of autumn would reach Christmas in a comparatively short time. But of course something was done about it: 1964 was a leap year with an extra day on February 29, and so the 00h 17m was in fact back on the 23rd. Since 1931, when it was 24d 00h 20m, this equinox has fallen regularly on the 23rd. It was 22d 23h 26m in 1968, but for some time to come three out of four September equinoxes will still be on the 23rd, which is convenient for the sidereal time estimate on page 30. The insertion of an extra day every fourth year goes back to Roman times and forms the Julian calendar.

Now, the year exceeds an exact number of days by 5h 48m 46s, and in four years this amounts to 23h 15m 4s. Putting in an extra day of 24h overcorrects by just about 45m in four years. In 100 years this adds up to $18\frac{3}{4}$h, so the century leap day is omitted, the year remaining 365 days instead of taking 366. This has cut $5\frac{1}{4}$h too much out of the century; in four centuries it has become 21h and the leap year day is put back. This is the Gregorian calendar (see Note 6, page 105) which we now use, in spite of an error of 3 hours in 400 years! The rule for a leap year is that the last two figures of the date shall be divisible by 4. The century year is not a leap year unless the first two figures are divisible by 4; thus 1900 was not a leap year, but 2000 will be.

3
The Celestial Sphere

CO-ORDINATE SYSTEMS

ASTRONOMICAL TRIANGLE

DIP AND PARALLAX

Books on mathematical astronomy nearly always begin with this chapter, and for users already fully accustomed to mathematical thought it is a very natural and logical starting point (see Note 7, page 105). This book is unorthodox in this respect, and readers will approach the conception of the celestial sphere already familiar with some of the terms used. It is rather like assembling a puzzle after some of its parts have previously been examined. They will be restated in their new setting, possibly from a different point of view, but before going on to the new setting it would be well to revise them in their former context. Here they are, with page references:

Latitude and longitude	3	Celestial longitude	13
Meridian	3	Celestial equator	22
Ecliptic and its plane	6	Sidereal time	23
Celestial pole	6	Hour angle	23
First Point of Aries	13	Equinox	32

The sky overhead looks like a dome on which we see the stars. A planetarium is a dome, a hemisphere, on which the star images are projected. Imagine the floor of the planetarium to be made of glass, below which is the other hemisphere: the whole thing would then be a model of the celestial sphere, the observer in the middle, representing the Earth, and the stars all around in every direction.

Study Fig 12. It shows the celestial sphere with the Earth in the middle; the Earth is marked with its axis, the equator, the observer's meridian and the direction of rotation. The intersections of the Earth's axis with the celestial sphere, ie the points in space vertically over the poles, are the *celestial poles* (NS). The line around the heavens vertically over the equator and everywhere 90° from the poles is the *celestial equator* (Q). The annual path of the Sun through the constellations is the *ecliptic* (C); the plane of the equator is inclined to the plane of the ecliptic by $23\frac{1}{2}°$ (approx.; ϵ, page 6). The motion of the Sun along the ecliptic is anticlockwise as seen from above, ie from left to right in the front of the sphere. The equator and ecliptic cut in two points, the *equinoctial points*. The one at which the Sun moves northwards in March is called the *Vernal Equinox*, and is also known as the *First Point of Aries* (although it is actually in Pisces); it is represented by the sign ♈. The other one, the Autumnal Equinox or the First Point of Libra, is marked ♎. This First Point of Aries has been used so far simply as an arbitrary zero of longitude and sidereal time; now it has a more definite meaning. When ♈ is in transit over the meridian the sidereal clock reads 0h. The star X will not transit until the

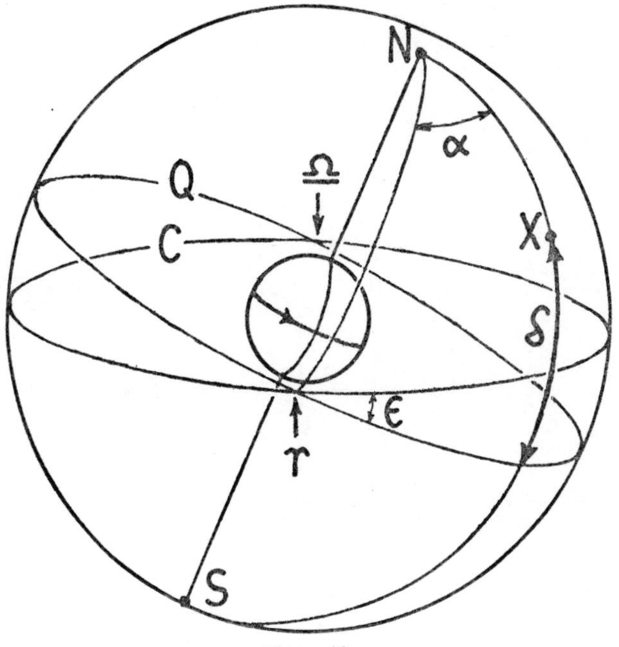

Figure 12

Earth has turned through the angle α. The time interval corresponding to this angle is the *Right Ascension* (symbol α, alpha) and is one co-ordinate used to fix a position on the celestial sphere. When X does reach the meridian, the reading of the sidereal clock will be the same as the R.A. of X. Thus the Right Ascension (R.A. or α) of a body is the sidereal time of its transit. The other co-ordinate required to fix X is its angular distance δ north (or south, when it is given a – sign) of the equator along a great circle through the poles; this is called *Declination* (δ, delta). In the observatory these two co-ordinates are measured with the transit circle mentioned at the beginning of the last chapter.

Turn to Fig 13. Imagine that you are in the planetarium facing the south horizon, and that the lines of R.A. and declination as well as the stars have been projected onto the dome; this usually can be done in a planetarium. The sidereal time is evidently 2h, for R.A. 2h is on the meridian (*a*). The star A, in R.A. 22h 30m, dec. +15°, passed the meridian 3½ hours ago, so the *local hour angle* (LHA, *b* in the figure) of the star is said to be 3h 30m. Hour

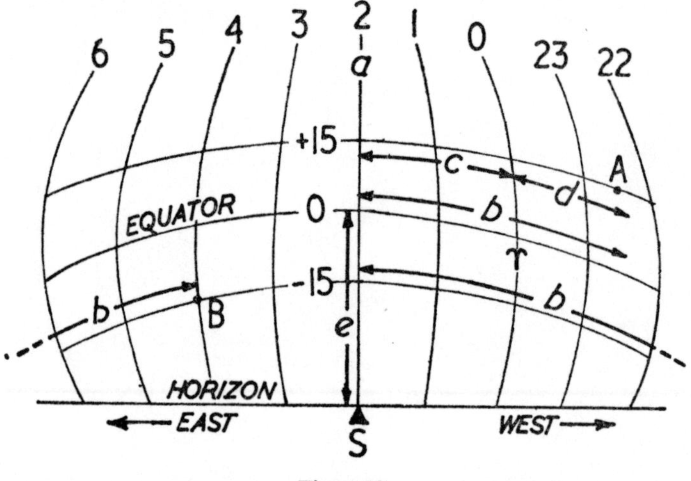

Figure 13

angles can also be expressed in degrees (page 26), and at 15° per hour it would be 52½°. Star B will not transit for another 2 hours; that is to say, 22 hours have elapsed since it last did transit, so its hour angle is 22h. The line of R.A. through the star, such as that through B, is called its *hour circle*. All objects on that line have the same hour angle, all will transit at the same time, and all have the same R.A. on star maps. The great circle marked NXS in Fig 12 is the hour circle of X. The measurement *c* in Fig 13 is the *hour angle of Aries* (H.A.♈); it is 2h in time units and was used in Chapter 2 as the definition of sidereal time. These hour angles change with time, but the measurement *d*, the *sidereal hour angle*

(SHA) does not and is an alternative co-ordinate to R.A. The sidereal hour angle of a star is the angle from the hour circle of Aries to that of the star, measured westward along the equator up to 360°; to find it, convert R.A. into degrees and subtract from 360.

Example:

$$\text{R.A. } 6\text{h } 20\text{m} = 6 \times 15 + \tfrac{1}{4} \times 20 = 95^\circ$$
$$\text{SHA} = 360 - 95 = 265^\circ$$

The measurement *e* is the *meridian altitude* of the equator.

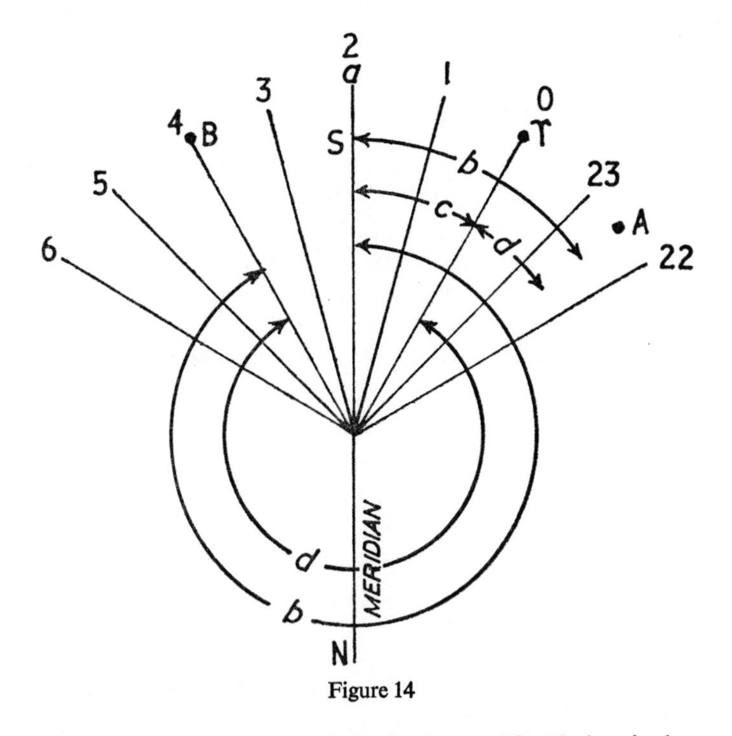

Figure 14

Fig 14 contains the same information as Fig 13, but is drawn in the plane of the equator. Compare it carefully with Table 3, and if necessary re-read the last paragraph with the new diagram—the lettering is the same. The last two lines, the relationships between LHA and the other factors, are important.

TABLE 3

Star	A	B
Right Ascension	22h 30m	04h 00m
Sidereal time (*a*)	02h 00m	02h 00m
Local hour angle (*b*)	03h 30m	22h 00m
Ditto, in degrees	52° 30′	330°
Hour angle of Aries (*c*)	30°	30°
Sidereal hour angle (*d*)	22° 30′	300°

Local hour angle in time = sidereal time − R.A. (+24h)*
Ditto, in degrees = H.A.♈ + SHA (−360°)*

* If necessary of course.

Fig 15 represents the plane of the meridian, still in the planetarium. The altitude of the pole (*a*) is equal to the latitude of the observer (page 112). The meridian altitude of the equator (*b*) is the complement of it, and therefore equals 90° − latitude. When the stars A and B are on the meridian, and only then, their altitudes (*c* and *d*) will be (90° − lat.) + dec., declination being negative

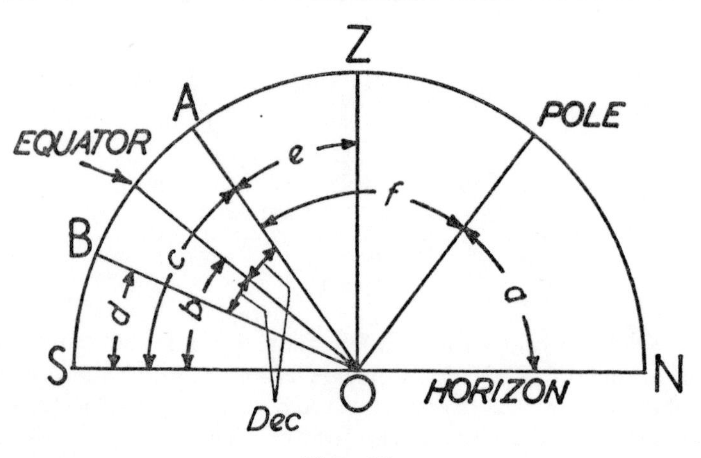

Figure 15

for B. In latitude 52°N their meridian altitudes would be $38° + 15°$ and $38° - 15°$ respectively, and less than that when not on the meridian. In general therefore

Meridian altitude =
co-latitude of the observer + declination of the object.

Z is zenith, the point overhead. The *zenith distance* (e) of an object at any time is the complement of the altitude at that time, because $c + e = 90°$. The complement of declination is called the *polar distance* (f); if this is less than the altitude of the pole the object will be circumpolar and will never set. Throughout this book 'transit' and 'crossing the meridian' means between the pole and the south point of the horizon. More precisely this is *upper culmination* or upper transit. A circumpolar star crosses the meridian again, between the pole and the north point, at lower culmination or transit. For observers in the southern hemisphere, north and south would be interchanged in this context.

Hour angle is important; the astronomer needs it for setting his telescopes, and the navigator uses it. It contains two elements: the positions of the stars on the celestial sphere, which can be tabulated more or less permanently, and a variable part which must be tabulated on a calendar basis. The 'permanent' part is not quite permanent. Owing to precession (page 6) the First Point of Aries moves westward along the ecliptic at a rate of 50″ per year. Thus star atlases carry the date or 'epoch' for which the co-ordinates were drawn and such information in the *Nautical Almanac* is revised regularly. This *precession of the equinoxes* also accounts for ♈ now being in the constellation Pisces. The calendar part is what the almanac must do, giving sidereal time or H.A.♈ for the meridian of Greenwich. But what if the observer is not on the meridian of Greenwich, longitude 0°? Fig 16 is the plane of the equator and O the centre of the Earth. The meridians through Greenwich, G, and two other places, A and B, are shown by radial lines. The hour angle at Greenwich (GHA) for the star S is a. As S is very distant compared with the radius of the Earth, it is permissible and convenient to measure all angles at the centre, so OS is the direction of the star for all observers. The local hour

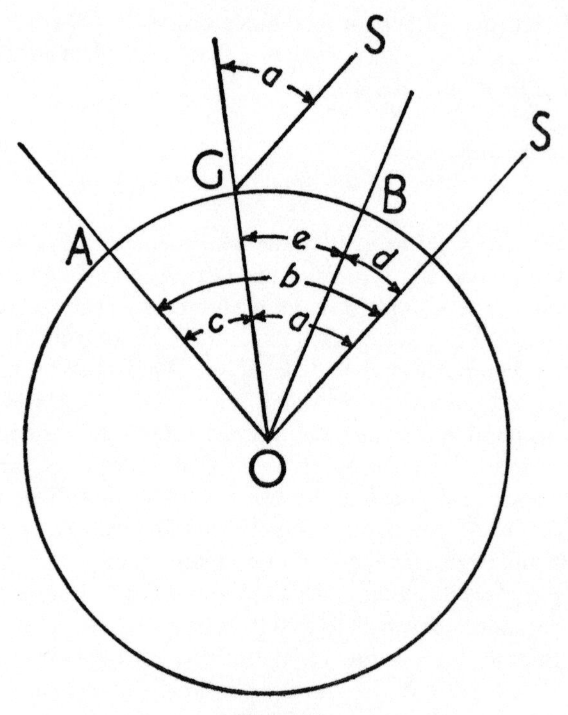

Figure 16

angle (LHA) at A (*b*) is the GHA plus the east longitude *c*; the LHA at B (*d*) is GHA minus the west longitude *e*. In general

$$\text{LHA} = \text{GHA} - \text{longitude},$$

west longitude being positive and east negative.

Example:

given that GHA = 90°
in long. 60°E LHA = 90 – (– 60) = 150°
in long. 110°W LHA = 90 – 110 + 360 = 340°

The reader should draw diagrams to illustrate these examples and any questions subsequently done; it is a safeguard—it is only too easy to get additions and subtractions the wrong way round if there is no picture of what you are doing.

Now let us link all these things together, Fig 17. The observer is at O. NPZS is his meridian, P being the pole and Z the zenith. X is a star. PZX is a spherical triangle made up of arcs of three great circles and measured in degrees. A great circle through the zenith and perpendicular to the horizon is *a vertical circle* (the meridian, for instance, is one), and it is along vertical circles that altitudes are measured. The *azimuth* of a body is the angle between the vertical circle through the pole and that through the body, measured through east up to 360°. This system of 0–360° from

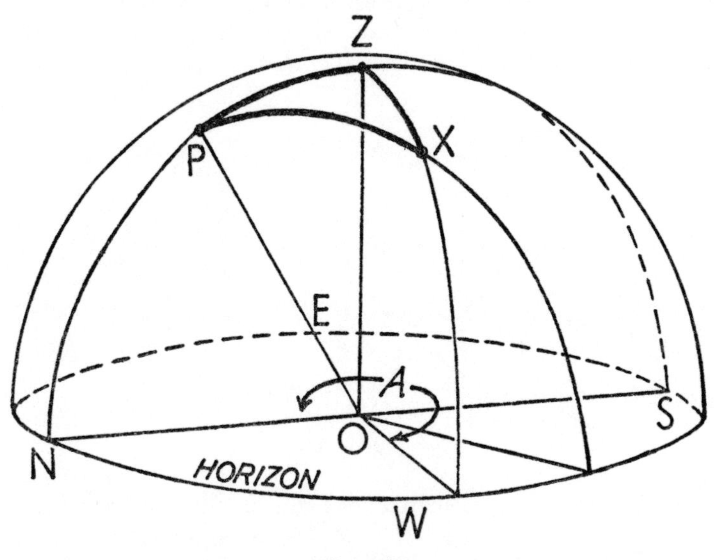

Figure 17

the north point, the same as true bearing, is not invariably followed. If you find the term elsewhere, check what system is being used. In this diagram the azimuth *A* is 360—angle PZX; if X were east of the meridian PZX would itself be the azimuth. As PZ is the hour circle on the meridian and PX that through the star

angle ZPX is the LHA = GHA – longitude,
angle ZXP is of no particular interest,
arc PZ is the zenith distance of P, = 90 – latitude,
arc PX is the polar distance of X, = 90 – declination,
arc ZX is the zenith distance of X, = 90 – altitude.

Thus the triangle contains two measurable quantities, altitude and azimuth; two tabulated quantities, GHA and declination; two positional quantities, latitude and longitude. The solution of this triangle by spherical trigonometry is the mathematical part of a navigator's work, and will be found in text books of navigation. The Sun, Moon and planets are not fixed on the celestial sphere, and this introduces another variable. In the *Nautical Almanac*, however, it has been included with the time element, and GHA for these bodies is tabulated, together with GHA♈, for short intervals of time throughout the year. Users of other almanacs must deduce their GHA from R.A. and sidereal time.

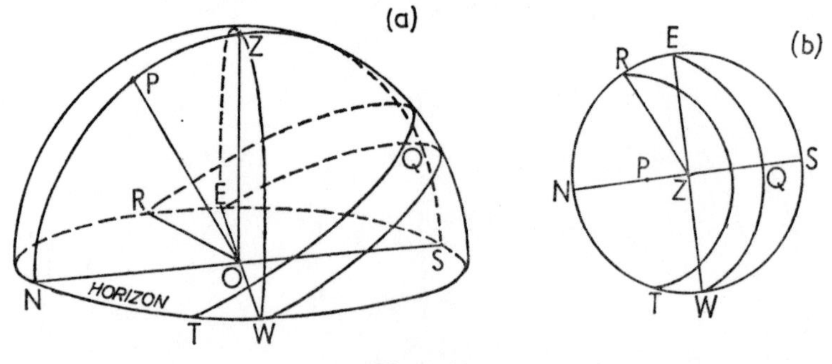

Figure 18

Fig 18A is the same celestial hemisphere, but showing different information. The vertical circle EZW is called the *prime vertical.* EQW is the familiar celestial equator, a plane perpendicular to OP. Notice that it meets the horizon at the points E and W; thus an object on the equator (such as the Sun at an equinox) will rise exactly east, be above the horizon for half a day, and set exactly west. An object with a positive declination will have a diurnal path parallel with the equator but north of it, meeting the horizon north of the prime vertical at R and T. The angle ROE is the *amplitude*, the angle along the horizon from the prime vertical to the point of rising or setting. It is of interest to the navigator and tables are available to him. The reader can imagine for himself the behaviour of an object with a negative declination. If the sphere were viewed from above the zenith it would look rather like Fig 18B, which is the conventional way of representing the same information (and lettering) as before.

The measurement of altitude, which a navigator does with a

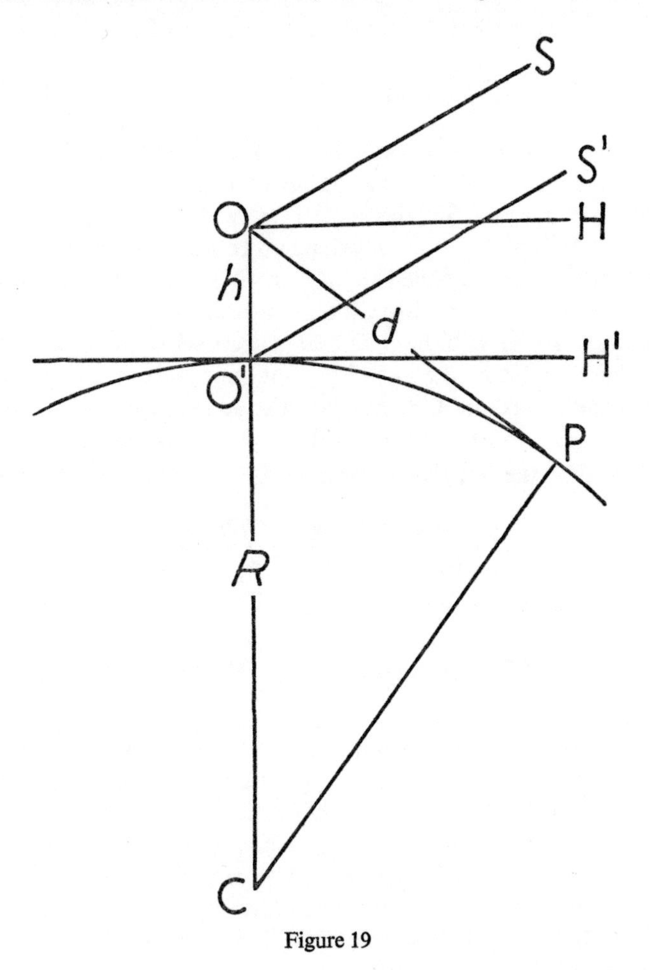

Figure 19

sextant, involves several other considerations. The first is the geographical altitude of the observer, Fig 19. The observer is at O, at a height *h* above the spherical surface. His horizon, instead of being at right angles to the radius as is O′H′, is the tangent OP, and its length d when h is small compared with the radius R of the Earth can be calculated from the relationship $d = \sqrt{2Rh}$.

Example: What is the distance of the horizon for an observer 50 metres above sea level, taking the radius of the Earth to be 6 370km?

$$d = \sqrt{2 \times 6\,370 \times 0{\cdot}05} = \sqrt{637} = 25{\cdot}2\text{km}.$$

This would apply to a sea horizon, for irregularities in a land surface would nullify a calculation of this kind. The observed altitude of the star S is the angle POS, which is greater than the altitude H′O′S′ (= HOS) obtained from sea level by the angle HOP. This angle is called the *dip of the horizon*, and allowance must be made for it when altitude is being used in navigation; tables are available to help. Sunset at sea level would occur when the Sun falls below the line O′H′, but the observer at O will still see it until it passes OP; his day will be lengthened. The angle POH = angle PCO′, and since h is very small compared with the radius CO′ the arc PO′ = the line OP = $\sqrt{2Rh}$. Then the dip

$$D = \frac{\text{arc}}{\text{radius}} = \frac{\sqrt{2Rh}}{R} = \sqrt{\frac{2h}{R}} \text{ radian.}$$

Let us try it on our 50m friend.

$$D = \sqrt{\frac{2 \times 0{\cdot}05}{6\,370}} = 0{\cdot}00396 \text{ radian} = \frac{0{\cdot}00396 \times 180}{3{\cdot}14} = 0^\circ{\cdot}23.$$

The corresponding time would be $0{\cdot}23 \times 4 = 0{\cdot}94$min; but as the path of the setting Sun is not ordinarily perpendicular to the horizon, the delay in sunset would be greater than this.

The second consideration is *refraction* in the atmosphere. This should have turned up qualitatively in your previous reading in astronomy or geography, and quantitatively it is rather involved as it is affected by atmospheric conditions as well as by the zenith distance of the object concerned. Very briefly, the image of the heavenly body is displaced towards the zenith by about ½ degree at the horizon (we can just see the whole of the Sun when it is supposed to have set), decreasing rapidly with altitude, and becoming zero at the zenith itself.

The third consideration applies in practice only to observation of the Moon. In discussing Fig 16 we transferred our viewpoint from the surface of the Earth to the centre, assuming that the objects concerned were so distant that GS and OS were parallel. The distance of the Moon is not enough to justify this assumption. In Fig 20, when the Moon M is observed from O, the zenith distance is ZOM, whereas if it were observed from C it would be the smaller angle ZCM. From the properties of triangles, ZOM = ZCM + OMC. The correction required is the angle OMC, which is called the parallax. Parallax is greatest when the Moon is on the horizon, O′; then the angle at the Moon

$$P = \frac{\text{arc CO}'}{\text{radius O}'\text{M}} = \frac{\text{radius of the Earth}}{\text{distance of the Moon}}$$

$$= \frac{6\,370}{384\,000} \times \frac{180}{3{\cdot}14} = 0^{\circ}{\cdot}95.$$

This quantity, known as *horizontal parallax*, will vary inversely as the distance of the Moon, so it is tabulated in the almanac. H.P. can be defined as the angle subtended at the Moon by the semi-diameter of the Earth or, as has now become possible, half the angular diameter of the Earth as seen from the Moon. Parallax other than on the horizon can be found by multiplying the H.P. by the sine of the zenith distance. Other bodies can have parallax too, but it is only for the Moon that it is important; the parallax of the Sun is about 8″·8 and that of the stars immeasurably small.

So far the co-ordinates used have been based on the equator, and for the last page or two on the horizon. For planetary studies, as in the next chapter, it is more convenient to work from the ecliptic. *Celestial longitude* (λ) has already turned up, and is measured eastward from ♈ along the ecliptic; *celestial latitude* (β) is the angular distance north (+) or south (–) of the ecliptic along a great circle through the object. Thus there are three systems used in this book: R.A. and declination, using hour circles intersecting at the poles; azimuth and altitude, using vertical circles intersecting at the zenith; longitude and latitude, using circles perpendicular to the ecliptic and intersecting at the poles of the ecliptic. These poles are situated in Draco and Dorado, and about them the

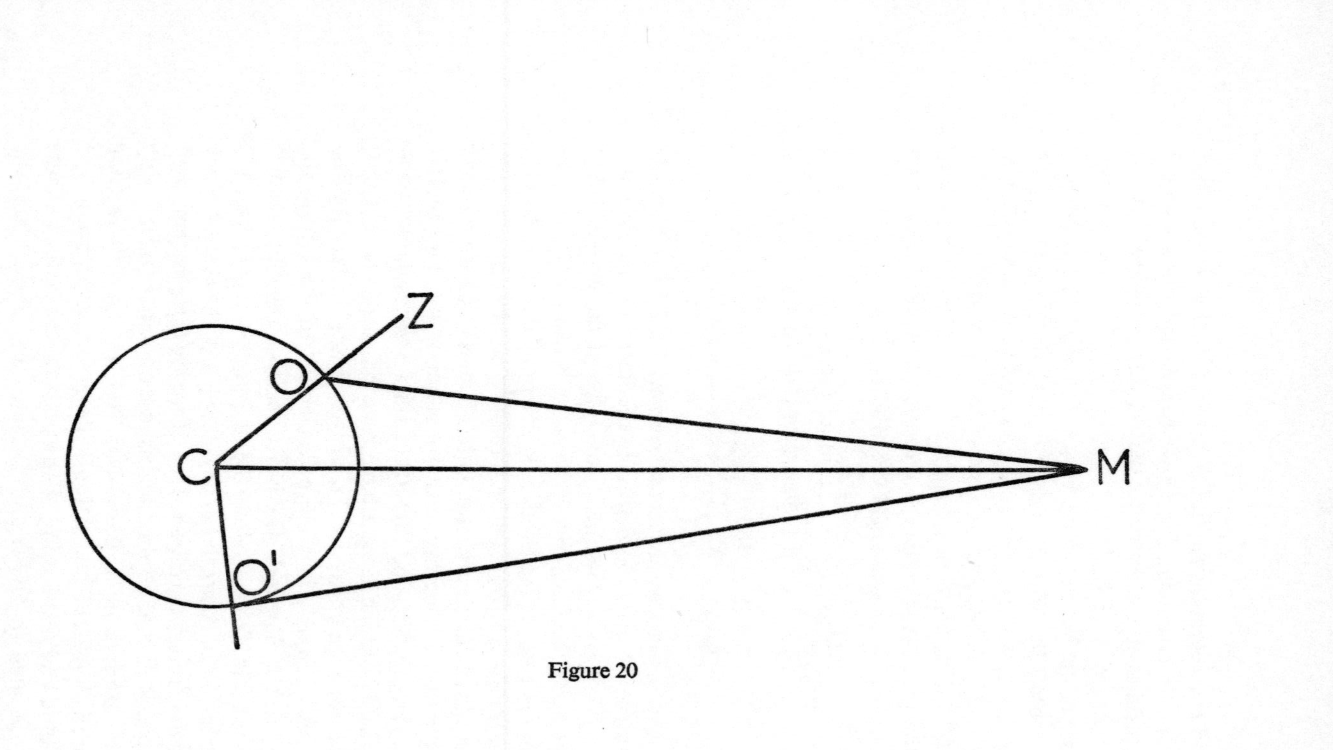

Figure 20

celestial poles make their precessional circuit in 26 000 years at a radius of 23½° (see Fig 3). The conversion from one system to another is a problem in spherical trigonometry outside our present scope. (See Davidson or Smart.)

Now, before going on to the exercises, read very carefully through this general example and if there is any step which you do not fully understand turn back to the appropriate page and revise it.

Example: The star Procyon, $\alpha = 07\text{h}\ 37\text{m}\ 44\text{s}$, $\delta = +05°18'11''$, is observed in transit at 23h 52m 29s LMT in $\theta = -21°01'48''$, $\phi = +52°13'05''$ on a day when the sidereal time at 0h UT was 07h 40m 15s. Find the error in the observer's clock and the meridian altitude of the transit.

			h	m	s
1	Longitude to time, p. 26	4m per 1° = 84m =	01	24	00
		4s per 1′ = 4s			04
		$\frac{1}{15}$s per 1″			03
			01	24	07
2	R.A. to Greenwich ST, pp. 37, 31	Observer's ST = R.A. =	07	37	44
		Greenwich ST earlier by	01	24	07
		Greenwich ST	06	13	37
3	ST to UT, p. 31	Sidereal time at 0h UT	07	40	15
		Sidereal interval (subt)	22	33	22
		less 10s per hour (220)		03	40
		Mean time interval = UT	22	29	42
4	UT to LMT, p. 28	add long. E	01	24	07
		LMT	23	53	49
5	Clock error	Reading of clock	23	52	29
		Error		−01	20
		Correction to be applied		+01	20
6	Meridian altitude, p. 41		90°	00′	00″
		Latitude	52	13	05
		Co-latitude (subt)	37	46	55
		add declination	+05	18	11
		Altitude required	43	05	06

Note that verification of Greenwich time need not be done at Greenwich, and that mean time and sidereal time do not involve observation of the mean sun or of ♈, which, having no material existence, are not observable.

4
The Solar System

APPARENT MOTIONS

PERIOD OF A PLANET

KEPLER'S LAWS

PLANETARY ORBITS

ORBIT OF A COMET

GRAVITATION

ARTIFICIAL SATELLITES

APPARENT MOTIONS

The reader will already be familiar with the composition of the solar system, and should also be aware of the apparent behaviour of its members in the night sky. The topic is quite fully treated in most descriptive books. Just as a precaution, however, we will have a quick look at apparent motions before going on to the numerical work. The orbits of Mercury and Venus lie within that of the Earth, and the remainder outside it; the orbit of an inner planet is shown in Fig 21A and that of an outer one in Fig 21B.

Owing to the motion of the Earth, the Sun apparently moves eastward along the ecliptic and carrying the orbit of Venus with it. We must ignore this, imagine the Earth to be at rest at E, and consider the effect of the motion of the faster-moving Venus relative to the Earth. When Venus is at 1 it has the same longitude as the Sun: *superior conjunction.* As it moves towards 2 it becomes visible to the east of the Sun, sets after it and is therefore an evening star, and almost the whole of the illuminated hemisphere is presented. At 3 the angular distance from the Sun is at its maximum of about 46°, and the visible planet is half illuminated: maximum *eastern elongation.* Between 3 and 4 a crescent phase is exhibited, and at 4 the longitude is again the same as the Sun's: *inferior conjunction.* From 1 to 4 the planet has been getting nearer, apparently larger, and brighter in spite of the decreasing phase, maximum brightness occurring soon after 3 when the elongation is about 40°. After 4 the planet is west of the Sun and rises before that body as a morning star; 5 is maximum *western elongation.* Mercury behaves in the same way, but the maximum elongation is only about half that of Venus, making the planet much more difficult to observe.

The outer planets have an eastward motion of their own, but we will ignore it and consider the effect of the motion of a faster Earth relative to a fixed planet at P. When the Earth is at 1 the planet is in conjunction. As the Earth moves towards 2, the planet becomes visible to the west of the Sun as a morning star, and its elongation increases all the time to 90° at 2 (*quadrature*) and

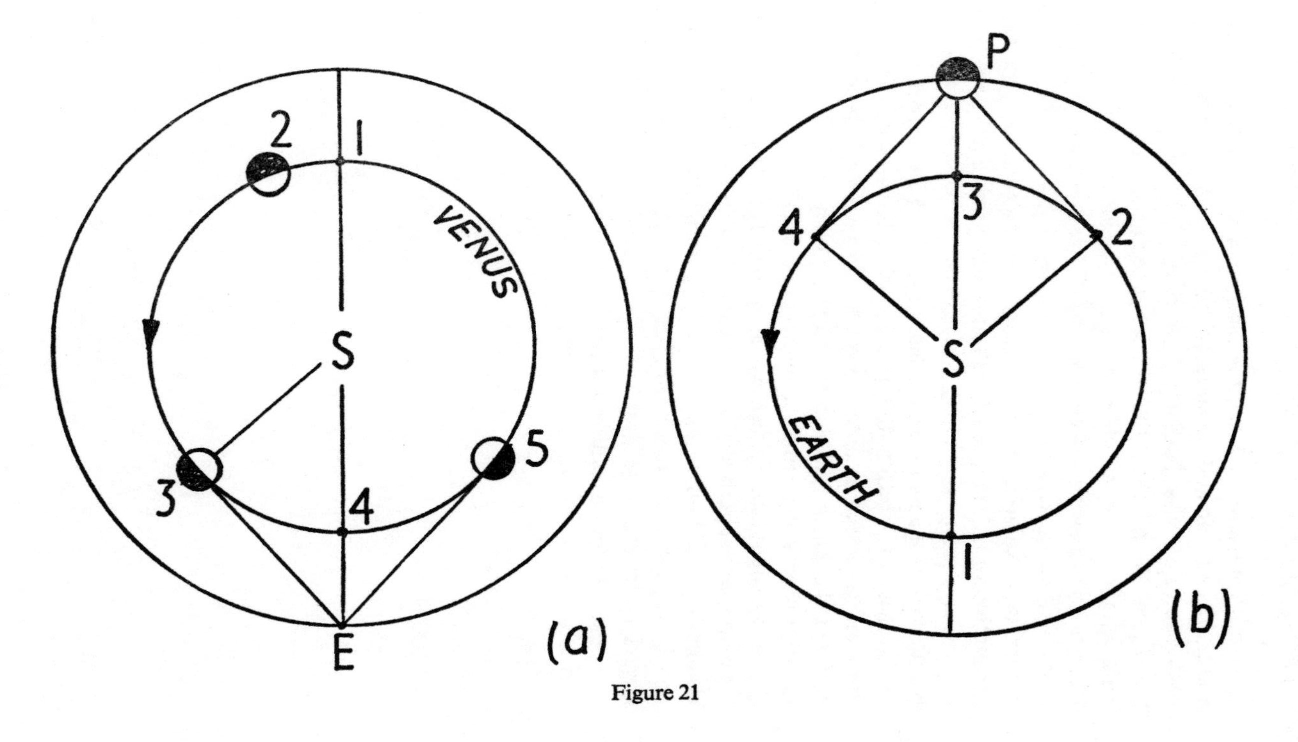

Figure 21

180° at 3 (*opposition*). Now, the motion of the Earth between 1 and 2 will cause an additional eastward motion of P relative to the background of stars, but after 2 the Earth will begin to overtake the planet. We all know that if a train begins to gain on another on a parallel track, an observer in the former will see the latter begin to go backwards; in the same way P has an apparent westerly motion impressed upon it. This soon more than counteracts its natural eastward motion, so for a time before and after opposition the motion of an outer planet is retrograde among the stars, with a *stationary point* at each end of this stage. At opposition the planet crosses the meridian at midnight; after that event it is seen to the east of the setting Sun (instead of to the west of the rising one) and is classed as an evening star. From 1 to 3 the distance between the Earth and P gets shorter, slowly at first and more rapidly towards opposition, so the brightness of the object increases. There is a slight phase effect, which is greatest in position 2, when the extreme west limb of P is unilluminated. This *defect of illumination* has a maximum of only about 12% of the diameter of Mars, and is negligible for the more distant planets. Maximum defect occurs again, of course, at 4.

Apparent motions will be further studied in the exercises at the end of the book, so back to the arithmetic.

THE PERIOD OF A PLANET

A planet has two periods, one relative to the Earth and another relative to the Sun. The interval between two successive similar conjunctions is the *synodic period*, and in that time an inner planet must gain 360° on the Earth, or the Earth gain 360° on an outer planet. Working only to the nearest day, the *sidereal periods*, once around the Sun measured against the stars, of the Earth and Venus are 365 and 225. Thus Venus moves

$$\frac{360}{225} \text{ degrees per day and the Earth } \frac{360}{365} \text{ per day.}$$

Therefore Venus gains $\frac{360}{225} - \frac{360}{365}$ degrees per day.

To gain 360° would take $360 \div \left(\frac{360}{225} - \frac{360}{365}\right) = \text{S days.}$

$$360 = S\left(\frac{360}{225} - \frac{360}{365}\right) \quad \text{or} \quad \frac{1}{S} = \frac{1}{225} - \frac{1}{365}$$

and

$$S = \frac{365 \times 225}{365 - 225} = 587 \text{ days.}$$

The actual value of the synodic period of Venus is 584 days. For the periods of the inner planets we have

$$\frac{1}{\text{synodic}} = \frac{1}{\text{sidereal}} - \frac{1}{\text{sidereal (Earth)}}.$$

The reader should prove for himself that for an outer planet

$$\frac{1}{\text{synodic}} = \frac{1}{\text{sidereal (Earth)}} - \frac{1}{\text{sidereal}}.$$

The synodic period, the interval between successive occurrences of the same Sun–Earth–planet configuration (or pattern) is the measurable one, and the sidereal period in the table books the derived one.

Example: Oppositions (i.e. transits nearest to midnight) of Mars occurred on 1967 April 15 and 1969 May 31. Find the sidereal period.

Days to		
	1967 May 1	16
	1968 May 1	366
	1969 May 1	365
	1969 May 31	30
Synodic period		777

$$\frac{1}{777} = \frac{1}{365} - \frac{1}{P}$$

$$\frac{1}{P} = \frac{1}{365} - \frac{1}{777}$$

$$P = \frac{777 \times 365}{412} = 688 \text{ days.}$$

KEPLER'S LAWS

The history of the gradual unravelling of the mystery of the planets from the earliest times to Kepler and Newton is a fascinating story and a 'must' for any student of astronomy. If you have not yet read it put it next on your book list. (In this connection attention is drawn to *Nuffield Physics Background Book: Astronomy*, to be published by Longman/Penguin early in 1972. It was specially written to show the development of a scientific theory in Year 5 (O level) of the Nuffield Science Teaching Project.) Here we go straight on to Kepler's conclusions (1609–1618), Fig 22.

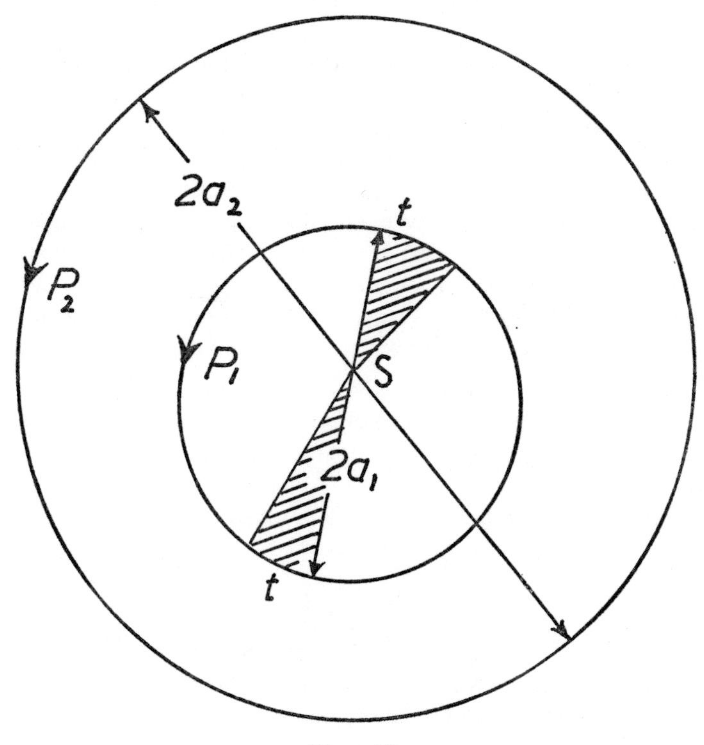

Figure 22

Law 1: The orbit of a planet is an ellipse having the Sun at one focus.

You have already drawn, from first principles, the orbits of the Earth and Moon (Exercises 15 and 16) and, assuming them to be ellipses, have measured their eccentricity. Readers who understood the note on page 104 can use the equation there quoted to test one or more points on these orbit drawings, which are supposed to have been preserved for further use. x and y are the co-ordinates of any point, measured parallel with and perpendicular to the major axis. a and b are the semi-major axis (already measured) and semi-minor axis (to be measured, and do not be surprised if it is almost the same as a), substitute in the equation and see how nearly it comes to 1.

Law 2: The motion of the planet is such that the radius vector from the Sun to the planet sweeps out equal areas in equal times.

The equal times are represented by t in the diagram and the equal areas are shaded. The areas are equal to $\frac{1}{2} \times$ length of arc $\times$ radius; apply this to your Moon orbit, which is already divided into sectors of $t = 2$ days. Work out several of the areas and see how nearly they are equal; do not expect perfection. It should be clear both from the figure and the formula that the longer the radius the shorter the arc, and consequently the less the velocity of the planet. This is the cause of the variable motion of the Sun in the ecliptic and of the Moon in its path, for while approaching perihelion a body will be accelerating, and slowing down when leaving it.

Law 3: The squares of the period times of the several planets are directly proportional to the cubes of their mean distances from the Sun.

Using the lettering of Fig 22, this becomes

$$\frac{P_1^2 \propto a_1^3}{P_2^2 \propto a_2^3} \quad \text{or} \quad \frac{P_1^2}{P_2^2} = \frac{a_1^3}{a_2^3}$$

where P is the sidereal period, the time taken for one revolution in the orbit, and a the semi-major axis of the ellipse which, as we know from Chapter 1, is the mean distance. Kepler did not know the actual mean distances; we can look them up in tables, but that is not allowed at this stage because the law has been involved in making the tables. Kepler did know the distance of certain planets relative to that of the Earth, that is on a scale of Earth-distance = 1, and in the case of Venus so do we. In Fig 21A the angle SE3 is known by observation to be 46°, and *if we assume* that the orbit of Venus is nearly a circle, the line of sight E3 will be a tangent to it and angle E3S = 90°. (In any argument it is wise to consider carefully what is a fact and what is assumed.) Then the ratio

$$\frac{\text{Venus distance S3}}{\text{Earth distance SE}} = \text{sine of the angle SE3} = \sin 46 = 0{\cdot}72.$$

Now, taking the ratio Earth/Venus and working to slide rule accuracy, we get:

$$\frac{a_1^3}{a_2^3} = \left(\frac{1{\cdot}00}{0{\cdot}72}\right)^3 = 1{\cdot}39^3 = 2{\cdot}69 \quad \text{and} \quad \frac{P_1^2}{P_2^2} = \left(\frac{365}{225}\right)^2 = 1{\cdot}62^2 = 2{\cdot}63.$$

The 365 days is derived from observations of the Sun, and the 225 calculated from the observed synodic period of Venus, so we have not used the law in any respect while trying to illustrate it.

Kepler's third law makes it possible to draw the whole system of planets on a scale of Earth-distance = 1, and it is the scale commonly used for solar system studies. Investigating planetary orbits and their laws from first principles, and preferably from personal observation, is the main theme of Tricker's book. The Earth-distance is called the *astronomical unit* (A.U.) and is of fundamental importance. If any one distance in the solar system can be measured in linear units, such as kilometres, then there would be a scale for the whole of it, but the A.U. itself cannot be measured directly. The horizontal parallax of the Moon (page 47) can be measured by, say, simultaneous observations of its position among the stars as seen from widely separated points on the Earth's surface, and from this the distance can be calculated. The parallax of the Sun cannot be found in this way; for one thing the angle is very small, and moreover there is no background of stars against which to measure it. The distance of one of the nearer planets can be so measured, and if the positions in their orbits of the Earth and the planet at the time are known, the scale will be determined and the A.U. can be calculated. Over the last three hundred years Mars and Eros have been used against the stars, and Venus against the disk of the Sun on several of the rare occasions when it was exactly between us and the Sun at inferior conjunction—an event called a 'transit of Venus'. A recent measurement was that of the distance of Venus by radar methods in 1964. The value of the astronomical unit is included with the accompanying table of planetary data.

PLANETARY ORBITS 1: JUNIOR

The orbits will be drawn as concentric circles and the only problem is to make them the correct radius (see Table 4). Choose any length for 1 astronomical unit, the guiding factor being the size

TABLE 4

ELEMENTS OF THE PLANETARY ORBITS

Name	1	2	3	4	5	6	7	8
Mercury	0·387	88d	077	0·206	048	7·0	048	4·09
Venus	0·723	225d	131	0·007	076	3·4	265	1·60
Earth	1·000	365d	102	0·017	not applicable		100	0·995
Eros*	1·46	642d	122	0·223	304	10·8	—	—
Mars	1·52	687d	336	0·093	049	1·8	013	0·524
Jupiter	5·20	11·9y	014	0·048	100	1·3	203	0·083
Saturn	9·54	29·5	092	0·056	113	2·5	043	0·033
Uranus	19·2	84	170	0·047	074	0·8	184	0·012
Neptune	30·1	165	044	0·009	131	1·8	239	0·006
Pluto†	39·8	251	223	0·25	110	17	195	0·004

1 Mean distance, *a*, in astronomical units. 1A.U. = 149 000 000km or 92 957 000 miles. Use 150×10^6 unless otherwise instructed.
2 Sidereal period, *P*, in days or years.
3 Longitude of the perihelion, ϖ, in degrees.
4 Eccentricity, *e*.
5 Longitude of the ascending node, ☊, in degrees.
6 Inclination to the plane of the ecliptic, *i*, in degrees.
7 Longitude of the planet on 1 Jan 1970. For the Earth this is approximately the same every year, but for the others a current almanac must be used. The accompanying exercises are for 1970.
8 Mean daily rate, *n*, in degrees.

* An asteroid, or minor planet.
† Elements for 1970; mean values are given in the rest of the table.

of your paper in relation to the largest orbit you want to draw. Suppose that it is to be Jupiter; the table of planetary orbits gives the distance to be 5·2A.U. If you choose 2cm to the unit, that would give 10·4cm radius, calling for paper not less than 25cm wide; the Earth orbit would be 2cm radius, of course. If you had that size of paper for the example in Fig 23 your Earth radius could be 10cm, and that of Venus would be $0{\cdot}723 \times 10 = 7{\cdot}23$cm. Draw the orbits; mark the Sun S at the centre; draw S♈ as the zero of longitude. The questions to be answered are these: if Venus crossed the meridian at 14h on April 10, and the telescope

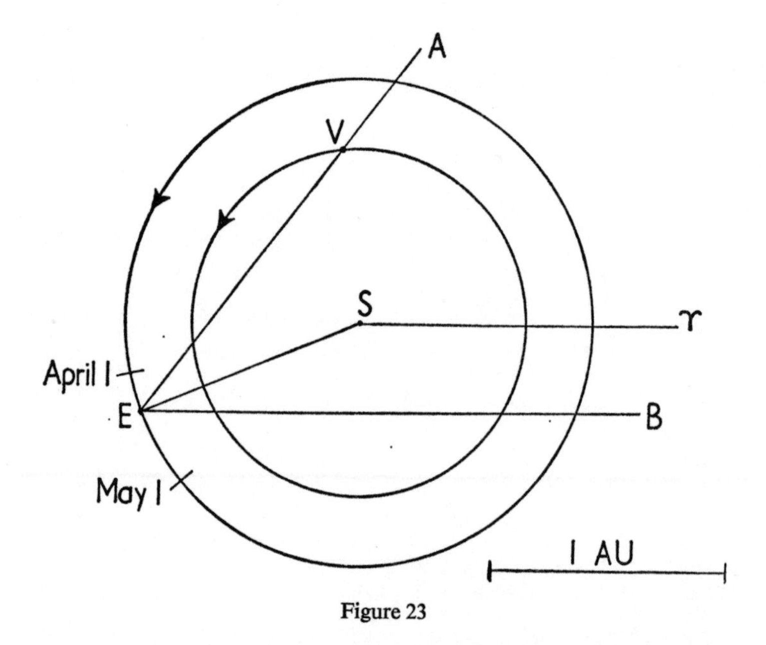

Figure 23

showed that it was not a crescent, find (i) the positions of the Earth and Venus in their orbits, (ii) the longitude of Venus as seen from the Earth, (iii) the constellation in which it would lie, and (iv) its distance from the Earth in kilometres.

(i) The table on page 13 shows that the geocentric longitudes of the Sun on April 1 and May 1 are 11° and 40° respectively, so the heliocentric longitudes of the Earth are $11 + 180°$ and $40 + 180°$. These are measured anti-clockwise from S♈, and are shown in Fig 23. April 10 would occur one-third of the way between April 1 and May 1, at E; join ES. Venus crossed the meridian two hours after the Sun, so it must be 30° to the east of it; make SEA = 30°. This cuts the orbit of Venus twice, but as the planet is not a crescent it cannot be at the nearer intersection. Therefore V is the position of Venus. For an outer planet there would be only one intersection.

(ii) Draw EB parallel with S♈. Then angle BEV is the longitude as seen from the Earth, 51° in this example.

(iii) Star maps do not normally show longitudes, but the adjacent table gives to the nearest degree the longitude at which the ecliptic crosses each hour of R.A. From this, long. 51° lies between R.A. 3h and 4h, say at 3h 20m. Reference to a star atlas shows that this is on the boundary between Aries and Taurus; and as the main planets are never very far from the ecliptic, that is where Venus will be.

(iv) Measure EV; use your scale to convert to A.U., which in Fig 23 comes to 1·35. 1A.U. = 150×10^6km, so the distance of Venus is $1{\cdot}35 \times 150 \times 10^6 = 202 \times 10^6$km.

TABLE 5

LONGITUDE ON THE ECLIPTIC AT EACH HOUR OF R.A.

R.A.	Long.	R.A.	Long.	R.A.	Long.
1h	16°	9h	133°	17h	256°
2	32	10	148	18	270
3	47	11	164	19	284
4	62	12	180	20	298
5	76	13	196	21	313
6	90	14	212	22	328
7	104	15	227	23	344
8	118	16	242	24	360

PLANETARY ORBITS 2: SENIOR

The orbits will be drawn as eccentric circles, and to enable this to be done a table of the elements of the planetary orbits is provided. It is recommended that the student make two large drawings on

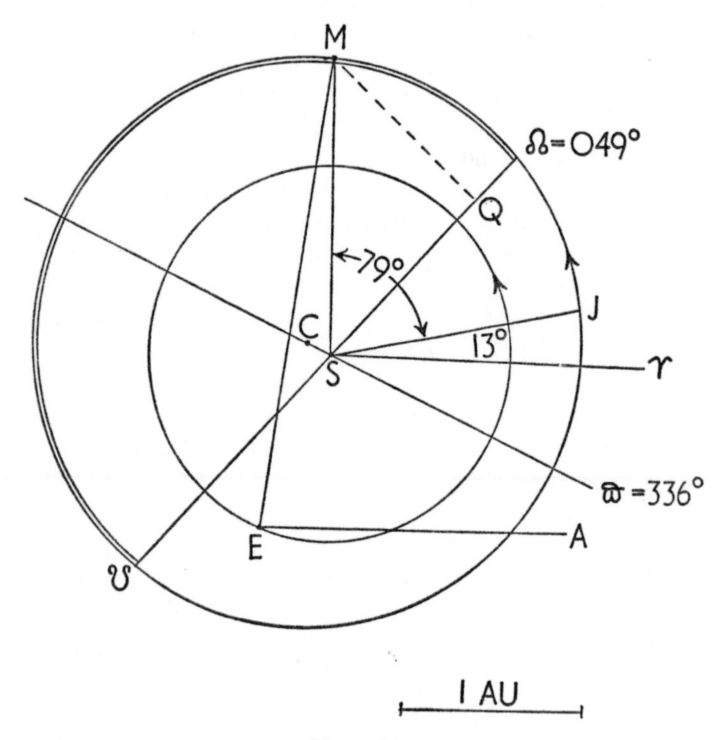

Figure 24

good-quality paper, one for Mercury, Venus, Earth and Mars, and the other on a smaller scale for the Earth and the four great planets—plus Pluto if space allows. Then if these drawings are completed in ink they can be used as the bases of the exercises instead of drawing fresh ones every time. As an example of the procedure Fig 24 has been drawn for the Earth and Mars, but it is too small to be truly to scale (the eccentricity has been exaggerated).

Mark the Sun, S, in the middle of a large sheet of paper and draw S♈ as zero longitude. From column 3 of the table find the longitude of the perihelion, $\varpi = 336°$; plot this angle and draw the major axis through S. Choose a scale to suit the paper. Suppose you took 10cm to the A.U.; then as a from column 1 is 1·52, the radius of the orbit of Mars will be $1{\cdot}52 \times 10 = 15{\cdot}2$cm, and the paper will have to be about 35cm wide. A larger scale is desirable. From column 4 the eccentricity is 0·093, and $CS = a \times 0{\cdot}093$ (page 9) measured on the axis away from the perihelion. With centre C and radius a draw the orbit. From column 5 take the longitude of the ascending node and insert ☊ at 49° and ☋ 180° further on. The orbit from ☊ anti-clockwise to ☋ is north of the plane of the ecliptic (and above the plane of the paper); this should be indicated by doubling or thickening the line. Now consider the situation on 1970 June 1. Column 7 of the table gives the longitude of the planet on Jan 1 to be 13°; insert J. Column 8 gives the daily rate to be 0·524 deg., and from Jan 1 to June 1 is 151 days. Thus the planet will have advanced from J by $0{\cdot}524 \times 151 = 79°$; insert M. This position will not be quite correct, for the daily motion is not uniform and we have used an average value, but it is near enough to locate the planet among the constellations (see Note 8, page 105). The orbit and position of the Earth are dealt with in the same way. Through E draw the line EA parallel with S♈ and measure the angle AEM. This is the geocentric longitude of Mars, 82°. Star maps do not normally mark longitudes, but the table on page 63 shows that long. 82° on the ecliptic, near which the planet would be, is just about R.A. 5h 30m; then a star map will show that Mars is just passing from Taurus into Gemini on June. 1 To find its distance from the Earth, measure EM in terms of your scale of A.U.; it is about 2·53, and as 1A.U. = 150×10^6km, the distance of Mars on that date was $2{\cdot}53 \times 150 \times 10^6 = 380 \times 10^6$km.

The planets do not stray very far from the ecliptic, but are normally not actually on it in lat. 0°, and their latitude can be estimated by a geometrical method, Fig 25. P is the position of the planet as so far determined in the plane of the ecliptic, P′ its real

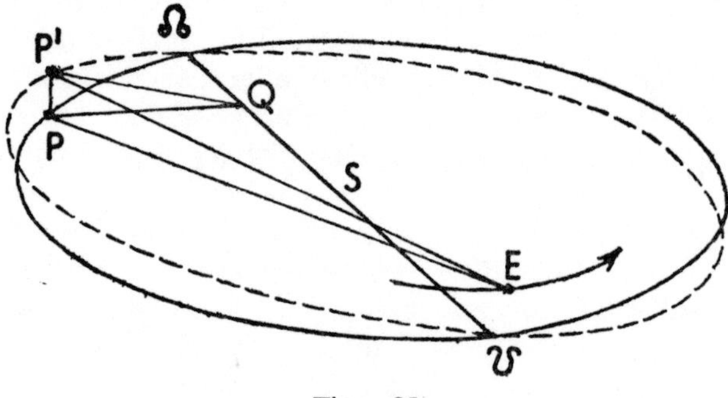

Figure 25

position in a north latitude and E the Earth. PQ is at right angles to the line of the nodes, so

$$\text{angle PQP}' = \text{inclination } i = \frac{PP'}{PQ} \text{ radian.}$$

Also $$\text{angle PEP}' = \text{latitude } \beta = \frac{PP'}{PE} \text{ radian.}$$

Dividing $$\frac{\beta}{i} = \frac{PP'}{PE} \div \frac{PP'}{PQ} = \frac{PQ}{PE}.$$

As the angles are now in ratio we can use degrees instead of radians without converting; i is in column 6 of table 4; PQ and PE can be measured from your orbit drawing. Applying it to Mars in Fig 24, the ratio MQ/ME $= 0{\cdot}43$, $i = 1^\circ{\cdot}8$ from the table, and $\beta = 0{\cdot}43 \times 1{\cdot}8 = 0{\cdot}78$. This is about $+47'$, $+$ because it is north of the ecliptic.

THE ORBIT OF A COMET

The orbit of a comet has elements generally similar to those of a planet, but the eccentricity is high and the resulting ellipses, far removed from the circle, are sometimes very long and narrow. For a short-period comet having a small orbit lying, for instance, within the orbit of Jupiter, the whole of it can be drawn, and method 2 (page 11) is convenient. For large orbits we need only the end near the Sun, and method 1 should be used. A long-period comet may have an orbit which is actually, or very nearly, a parabola, and in this case the element a becomes useless and is not given. Instead the perihelion distance q is quoted, $e = 1$, and ellipse method 1 must be used for drawing it.

As an example, Fig 26 shows the orbit of Comet 1927 VI, for which the elements (in the order in which we shall use them) are:

$$☊ = 66° \quad \omega = 210° \quad a = 4{\cdot}9 \quad e = 0{\cdot}76$$

$$T = 1970 \text{ Nov } 17 \quad P = 10{\cdot}9 \text{ years} \quad i = 11°$$

S is the Sun. For reasons which will appear later, it is convenient to start with the line joining the nodes, drawn horizontally through S with the ascending node on the right; then measure ☊ $= 66°$ clockwise to locate the usual S♈. For a planet ϖ was the longitude of the perihelion measured from ♈; ω for a comet is the *argument of the perihelion* measured from ☊ in the direction of motion, so take 210° anti-clockwise from ☊ to locate Sω. This is the axis of the ellipse. Choose a suitable scale of astronomical units and so express a in centimetres or inches. Use a and e to draw the ellipse; method 1 was used, but 2 could be. On the same scale draw the orbit of the Earth and any other planets you want; concentric circles will do. T is the date of perihelion passage; deduce the position of the Earth in the usual way, and insert the Earth and the comet for the date T. P is the period of the comet, 10·9 years for a complete circuit. The inclination i of the plane of the orbit to that of the ecliptic has so far been ignored, but in this case, being only 11°, it does not make much difference to a representation on paper. The ephemeris of a comet, as issued by the British Astronomical Association, gives the distances of the comet from the Earth (Δ) and from the Sun (r). The latter can be used, with your compasses or dividers, to mark the positions of the comet in its orbit on the various dates.

Many comets have high inclinations, and cannot be shown

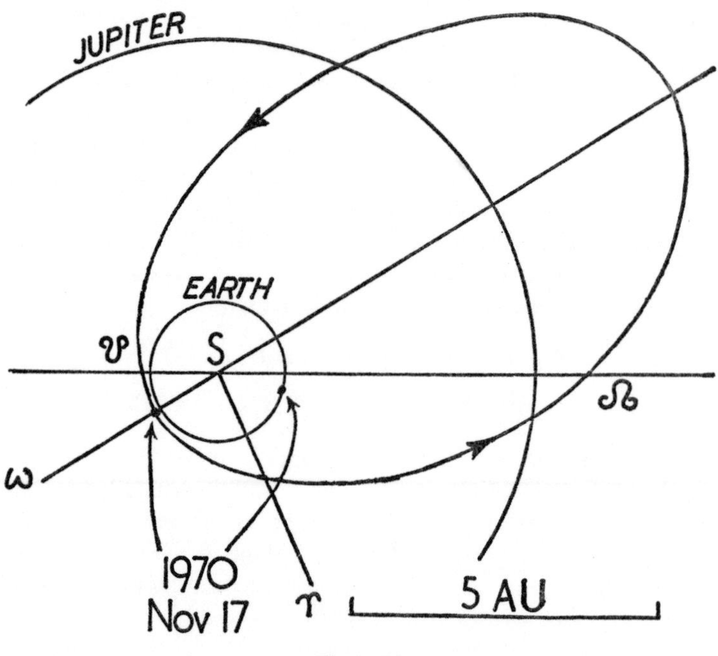

Figure 26

satisfactorily in the ecliptic plane; they must be shown three-dimensionally. Instead of paper you need two thin cards of the same width. The node line (with which the drawings start) is where the two planes concerned intersect; draw this, and the exact position of the Sun, on both cards. On the 'ecliptic' card draw S♈ and any planet orbits you want, and cut a narrow slot exactly half-way across the node line on the ascending node side. On the 'comet' card draw Sω and the comet orbit, and cut a slot in the node line from the descending node side. The two cards can now be interlocked as shown in Fig 27A; cut out a cardboard wedge (27B) to hold them at the correct angle i.

As angles of inclination get larger the comet card gets steeper, and if i passes 90° you will have something like Fig 27C. Your comet orbit is now disappearing in the sandwich, and ought to be on the other side of the card. Thus when, right at the start, you notice $i > 90°$, the comet card (including the slot) must be prepared as a 'mirror image' by putting the ascending node on the left. It is

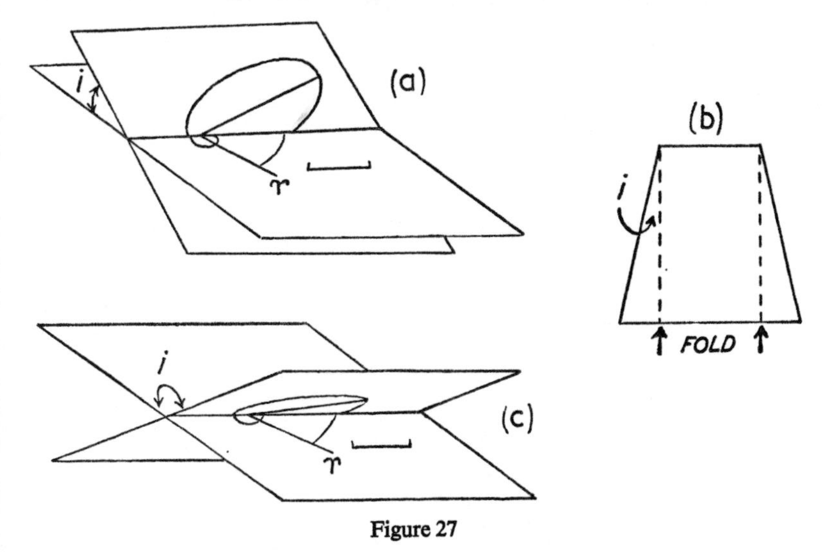

Figure 27

wise in these circumstances to put the Sun at the mid-point of the node line, or there may be difficulty in getting correct register when the cards are put together. The angle of the wedge will now be $180° - i$. Notice that in Fig 27A the direct-moving (anti-clockwise) comet passes to the north of the ecliptic card at the ascending node. So it does when you have assembled 27C, but as ☊ is now on the left when viewed from above the motion is clockwise—it is a retrograde comet. Inclination greater than 90° is the convention for indicating retrograde motion.

A meteor stream has an orbit similar to that of a comet, so we will close this section with a little problem about meteors. The elements of a certain meteor stream are: ☊ $= 233°$, $\omega = 179°$, $q = 0{\cdot}99$, $i = 163°$ ($e = 0{\cdot}91$). Estimate the date of the encounter and the direction from which the meteors should come.

Draw the node line, Fig 28, and measure 233° clockwise from ☊ to locate S♈ (as we are working on one sheet, there is no need

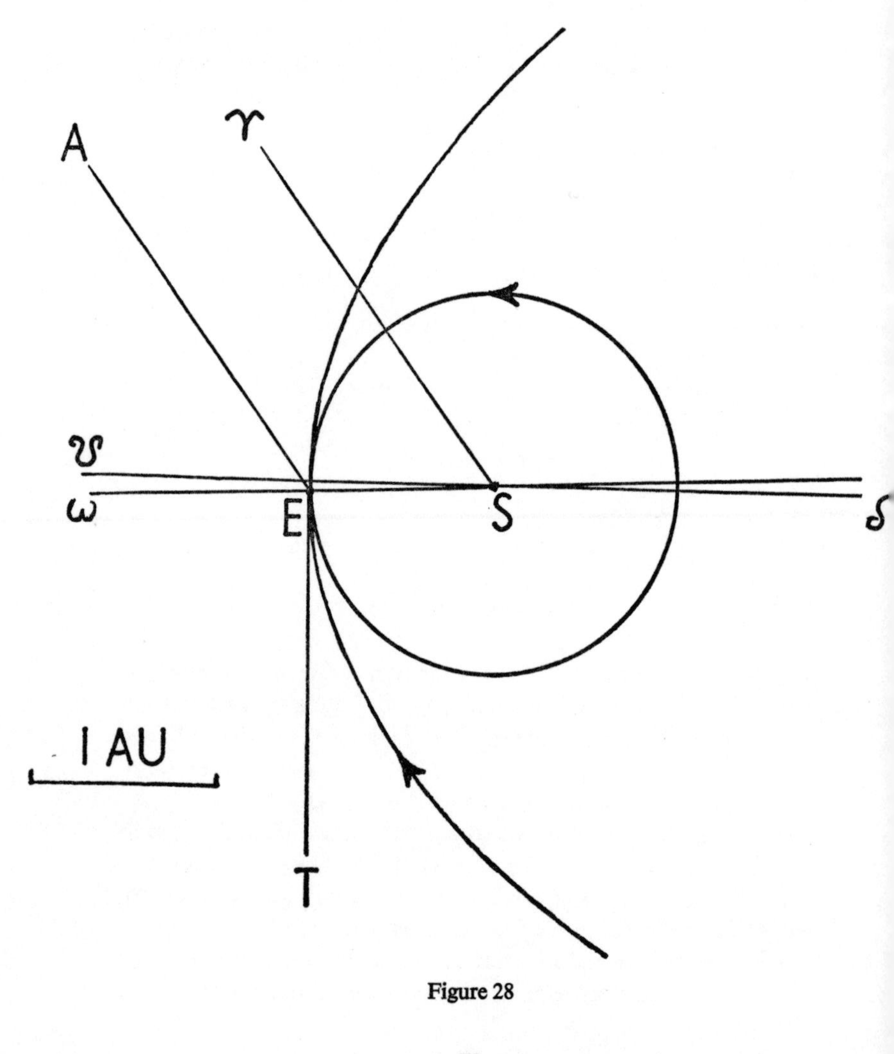

Figure 28

to use the mirror image). The value of i shows that the motion is retrograde, so in measuring $\omega = 179°$ from ☊ in the direction of motion it will be clockwise to fix the line Sω. When e is as high as this the curve near the focus is hardly distinguishable from a parabola for which $e = 1$, so the orbit can be drawn as such using ellipse method 1. When the orbit of the Earth is drawn it can be seen that it almost touches that of the meteors at E, the perihelion, so the shower would be expected when the Earth were there. The heliocentric longitude of E = ♈SE = 56°; geocentric longitude of the Sun on that day = 56° + 180° = 236°; the table on page 13 shows that to occur about November 18. The Earth and the meteors will be meeting almost head-on along the tangent TE. Draw EA parallel with S♈; then AET = long. of T = 145°. Reference to page 63 shows that this is not quite R.A. 10h, and as the meteors are approaching their descending node they would be a few degrees north of the ecliptic. Now look at a star atlas and see where the radiant is; an approximate position, of course.

GRAVITATION

Every material body attracts every other with a force which we call gravity, and their behaviour is the resultant of them all. When we realise how many astronomical bodies there are, even in the solar system, it is obvious that the subject is very complex as well as being of fundamental importance. All that will be attempted in this introductory book is to remind the reader of some elementary physics and consider a few astronomical examples of it. The physics can be amplified from any good 'O level' text book. In these examples we shall assume that the dominant force is the only one, something that the mathematical astronomer cannot do—he must incorporate the 'perturbations' caused by neighbouring bodies. The results will be approximations and the methods not necessarily those used in the profession. Gravitation is associated with the name of Newton; his work was published in 1687 but had been done years earlier—see the historical book which you are going to read next.

We begin with his *laws of motion.*

Law 1: Every body continues in its state of rest or of uniform motion in a straight line unless acted upon by a force.

We have already noticed that astronomical bodies are not at rest, and that their motions are neither uniform nor straight; they are subject to forces.

Law 2: The rate of change of momentum of a body is directly proportional to the applied force, and takes place in the direction of that force.

Momentum, in common speech, conveys the difficulty in stopping something. In mechanics it is the product mass × velocity, so a rate of change of momentum is mass × rate of change in velocity. Rate of change in velocity is acceleration, positive when velocity is increasing and negative when decreasing. The specification of a motor car may state that 60mph can be attained in 15 seconds, a rate of change or acceleration of 4mph per second.

Note that time units have come in twice. Any freely falling (no or negligible air resistance) body near the Earth's surface increases its velocity by 9·81 metres per second in each second of its fall, so the *acceleration due to gravity* (g) is 9·81 metres per second per second, written 9·81m s^{-2}. Force, then, is directly proportional to mass × acceleration, and units of force have been defined in such a way that

$$\text{Force} = \text{mass} \times \text{acceleration}.$$

If the mass be 1kg and the resulting acceleration 1m s^{-2}, the force being applied is by definition 1 newton, the SI unit of force. (Older physics books use cgs units; the unit of force is the dyne, which will give a mass of 1g, an acceleration of 1cm s^{-2}.)

Law 3: To every action there is an equal and opposite reaction.

This means that forces work in both directions; it is quite possible for a cork to pull the corkscrew out of its handle instead of the corkscrew pulling the cork out of the bottle. The attraction of the Earth for the Moon maintains the latter in orbit, but at the same time the attraction of the Moon for the Earth deviates it from a uniform ellipse around the Sun, as was mentioned in Chapter 1.

All freely falling bodies have the same acceleration, g; therefore the force of gravity on each must be proportional to its mass. As forces act both ways, it may be presumed that the forces are proportional to the Earth's mass also, and that the corresponding acceleration at the Moon's surface would be less—which astronauts' tales confirm. The force is likely to become less with increasing distance, so let us compare, as Newton did, according to tradition, in 1666, the familiar g with the acceleration of the Moon. (Usually not in 'O level' text books, though the formal statement (which follows) of the law of gravitation probably is.) When you swing a weight around on the end of a string you can feel a force along the string. From law 3 we know, therefore, that the string must be exerting a force on the weight, also along the string. This force must cause an acceleration, and it is this acceleration towards the centre which causes the weight to move in a circular path—a change in velocity can be a change in direction, not necessarily a change in the distance covered in unit time.

It is shown in books on mechanics that this acceleration is given by the relationship $a = v^2/r$, where v is the linear velocity in a circular orbit and r the radius of that orbit. Now the linear velocity of the Moon is

$$\frac{\text{circumference in metres}}{\text{sidereal period in seconds}} = \frac{2 \times 3{\cdot}14 \times 384 \times 10^6}{27{\cdot}3 \times 24 \times 3\,600} \text{m s}^{-1}.$$

Hence

$$a = \frac{v^2}{r} = \frac{4 \times 3{\cdot}14^2 \times 384 \times 10^6}{(27{\cdot}3 \times 24 \times 3\,600)^2} = 0{\cdot}00272 \text{m s}^{-2}.$$

This is less than g at the Earth's surface by the factor

$$\frac{0{\cdot}00272}{9{\cdot}81} = \frac{1}{3\,603}.$$

A uniform sphere, which we are assuming the Earth to be, acts gravitationally towards external bodies as if all its mass were at the centre, and as the radius of the Earth is 6 370km the distance of the Moon is greater than that of the surface by the factor $384\,000/6\,370 = 60{\cdot}28$. Evidently the force decreases much more rapidly than the distance increases. Suppose we square the distances: $384\,000^2/6\,370^2 = 60{\cdot}28^2 = 3\,634$. Considering the assumptions and averaging which have been used, the agreement is good enough to accept; the force is inversely proportional to the square of the distance. We can now state Newton's *Law of Gravitation:* the force of attraction between two bodies is directly proportional to the product of their masses, and inversely proportional to the square of the distance between them.

In mathematica lform it can be written

$$F \propto \frac{MM'}{D^2} \quad \text{or} \quad F = \frac{GMM'}{D^2}$$

where G, not to be confused with g, is the *general gravitational constant.* Its value, as determined by laboratory experiments, is $6{\cdot}67 \times 10^{-11}$ SI units ($6{\cdot}67 \times 10^{-8}$cgs).

Now we have a means of calculating the mass of the Earth. A mass of 1kg has an acceleration of $9{\cdot}81\text{m s}^{-2}$, so from $F = ma$

the force acting upon it is 9·81 newtons. Substituting in the gravitation formula we get

$$9{\cdot}81 = \frac{6{\cdot}67 \times 10^{-11} \times M \times 1}{6\,370\,000^2},$$

$$M = \frac{9{\cdot}81 \times 6\,370\,000^2}{6{\cdot}67 \times 10^{-11}} = 5{\cdot}97 \times 10^{24}\text{kg} = 5{\cdot}97 \times 10^{21} \text{ tonnes},$$

which is just about 6×10^{21} ordinary British tons.

Newton's law of gravitation can be used to explain Kepler's laws. Only the third law will be considered here, as it provides some helpful equations. Let M be the mass of the central body and M' the orbiting one, then

$$F = \frac{GMM'}{D^2} \quad \text{and} \quad a = \frac{F}{M'} = \frac{GM}{D^2}.$$

But

$$a = \frac{v^2}{D} = \left(\frac{4\pi D}{T}\right)^2 \times \frac{1}{D} = \frac{4\pi^2 D}{T^2}.$$

Hence

$$\frac{GM}{D^2} = \frac{4\pi^2 D}{T^2} \quad \text{or} \quad \frac{GM}{4\pi^2} = \frac{D^3}{T^2}.$$

As G, $4\pi^2$ and M are constant for any one central body, so is D^3/T^2, which is the third law.

We now have a means of finding the mass of the Sun. Let S and E be the masses of the Sun and Earth, D and d the distances of the Sun and Moon, and T and t the corresponding periods. Rearranging the last equation and substituting S and E for M we get: for Sun–Earth

$$S = \frac{4\pi^2 D^3}{GT^2},$$

for Earth–Moon

$$E = \frac{4\pi^2 d^3}{Gt^2},$$

dividing we get

$$\frac{S}{E} = \frac{D^3}{T^2} \times \frac{t^2}{d^3} \quad \text{or} \quad \left(\frac{D}{d}\right)^3 \times \left(\frac{t}{T}\right)^2.$$

Inserting the figures

$$\frac{S}{E}=\left(\frac{150\times 10^6}{384\times 10^3}\right)^3\times\left(\frac{27{\cdot}3}{365}\right)^2=333\ 000.$$

Thus the mass of the sun is 333 000 times that of the Earth, and it can be converted into kilograms if required.

Take another example of this useful process. Triton, the principal satellite of Neptune, is 0·0024A.U. from the planet and has a period of 5·88 days. Find the mass of Neptune (data for Neptune on page 61).

$$\frac{\text{Neptune–Triton}}{\text{Sun–Neptune}}=\frac{N}{S}=\frac{d^3\times T^2}{t^2\times D^3}$$

$$=\left(\frac{0{\cdot}0024}{30{\cdot}1}\right)^3\times\left(\frac{165\times 365}{5{\cdot}88}\right)^2=0{\cdot}000\ 053.$$

Thus Neptune is $5{\cdot}3\times 10^{-5}$ compared with the Sun, and in terms of the Earth that gives $(5{\cdot}3\times 10^{-5})\times(3{\cdot}3\times 10^5)=17{\cdot}5$.

We have assumed so far that the orbiting object is moving around an immovable centre, which it is not, but only very nearly so when its mass is insignificant compared with that at the centre. This certainly will be the case for the Earth and Sun, but not for the Earth and Moon. The Earth's attraction for the less massive Moon puts it into a large orbit; the Moon's attraction for the Earth puts it into a small one (radius 4 800km). A more correct relationship is

$$\frac{G(M+M')}{4\pi^2}=\frac{D^3}{T^2}$$

for any pair of orbiting bodies. Let

$$\frac{G(m+m')}{4\pi^2}=\frac{d^3}{t^2}$$

apply to the Sun-Earth system; dividing

$$\frac{M+M'}{m+m'}=\frac{D^3\,t^2}{d^3\,T^2}.$$

Similar quantities must be in similar units, of course; then the process of dividing removes all units (without introducing any multiplying factors such as years to days) and leaves a purely numerical expression. (Thus physics students familiar with the theory of dimensions need not worry, as some of the writers have done, about the apparent impossibility of the final equation.)

Take the Sun as the unit of mass: $m = 1$, m' is negligible
take the A.U. as the unit of distance: $d = 1$
take the year as the unit of time: $t = 1$
and the expression becomes $M + M' = \frac{D^3}{T^2}$

which is a general one; it applies not only in the solar system, but takes us to the stars, and will turn up again in the next chapter.

ARTIFICIAL SATELLITES

Nearly three centuries ago Newton showed that if a gun be set on a very high mountain outside the atmosphere and a shot fired parallel with the surface at a sufficiently high velocity, then it would go right around the Earth and be what we now call an *artificial satellite.* Of course there is no such mountain, and there is no such gun, but using modern rocketry the result was achieved in 1957, and artificial satellites have now become almost common-place. The details will be found in the 'space books'—works on *astronautics*—and all that is appropriate here is a quick look at the principle.

If a vehicle at a distance d from the Earth's centre be projected tangentially with a velocity v, then if also the acceleration due to gravity at that distance be v^2/d it will go into a circular orbit (page 74). But the acceleration is GM/d^2 (page 75), so

$$\frac{v^2}{d}=\frac{GM}{d^2} \quad \text{or} \quad v^2=\frac{GM}{d}.$$

Example: If the altitude were 630km, $d = 7\,000$km or 7×10^6m, then

$$v^2=\frac{6{\cdot}67\times 10^{-11}\times 5{\cdot}97\times 10^{24}}{7\times 10^6}=\frac{6{\cdot}67\times 5{\cdot}97\times 10^7}{7},$$

$$v = 7\,540\text{m s}^{-1} = 7{\cdot}54\text{km s}^{-1}.$$

This is the *circular velocity* at an altitude of 630km; near the Earth's surface it would be a little greater because d is smaller. As an alternative, which the reader can try for himself if he likes, apply Kepler's third law in the form

$$\frac{\text{period}^2}{\text{distance}^3}\text{ for satellite}=\frac{\text{period}^2}{\text{distance}^3}\text{ for Moon.}$$

This will give the period of the satellite, and by dividing it into the circumference of the orbit the velocity can be obtained.

Suppose that the velocity of projection were a little less than the circular one. Then the acceleration due to gravity would be greater than that necessary to maintain the circular motion; the satellite would therefore be drawn inwards into an elliptical orbit taking it nearer to the Earth on the far side—the point of projection would be apogee. If this applied to our example, an altitude of 630km is hardly outside the atmosphere, so the perigee is certainly not. This means that air resistance will check the velocity still further and generate heat. The object will tend to spiral inwards, getting hotter, and eventually to incandescence and, unless re-entry has been specially provided for, to disintegration.

If, on the other hand, the velocity of projection were greater than the circular value, gravity would be insufficient to hold the vehicle to a circle. An elliptical orbit would again result, but with the point of projection as perigee, and the higher the velocity the more remote apogee would be, perhaps as far as the orbit of the Moon. A rendezvous with the Moon itself would take us outside the limitation laid down earlier in the chapter—that we shall consider only one controlling gravitational force. Readers of the Apollo story will realise the importance to that achievement of the gravitational field of the Moon.

The greater the velocity of projection the further off the apogee and the more remote the far focus of the ellipse, until the distance of the latter becomes infinity and the ellipse has turned into a parabola. This is an open curve: the vehicle will never return. The parabolic velocity of projection is also called the *velocity of escape*, the velocity at or above which a body will escape from the bond of the Earth's gravity. Text books on physics show that it can be calculated from the relation $v^2 = 2GM/d$.

Applying it to the previous example (starting altitude 630km):

$$v^2 = \frac{2 \times 6{\cdot}67 \times 10^{-11} \times 5{\cdot}97 \times 10^{24}}{7 \times 10^6} = \frac{2 \times 6{\cdot}67 \times 5{\cdot}97 \times 10^7}{7},$$

$$v = 10\,670\text{m s}^{-1} = 10{\cdot}7\text{km s}^{-1}.$$

If the acceleration due to gravity, g, is known the formula can be made even simpler, for $g = GM/d^2$. Thus $GM/d = dg$ and so $v^2 = 2dg$.

At the Earth's surface $g = 9{\cdot}81\text{m s}^{-2}$ and $d = 6\ 370\text{km}$, so

$$v^2 = 2 \times 6\ 370\ 000 \times 9{\cdot}81$$
$$v = 11\ 180\text{m s}^{-1} = 11{\cdot}2\text{km s}^{-1}.$$

Notice that this is greater than the value at an altitude of 630km, so it requires less energy to despatch an interplanetary probe from a 'parking orbit' than from the surface of the Earth.

5
Stellar Topics

STAR CHARTS

STELLAR MAGNITUDE

STELLAR PARALLAX

ABSOLUTE MAGNITUDE

MASS OF BINARY STAR

DOPPLER EFFECT

This is not a chapter in the sense that there is a consistent theme running through it. It is really a miscellaneous collection of odds and ends largely concerned with stars. As students and amateurs need to use star charts we will begin with these, and assume that the reader is going to draw some as well as use them.

STAR CHARTS

As in geography there is the problem of projection, that of depicting a part of the celestial sphere on a flat sheet of paper, a process that inevitably involves some distortion. As with geographical atlases, star atlases use different systems according to the areas which they are trying to show. We shall use only two.

The middle heavens can be shown quite adequately with rectangular co-ordinates, illustrated in Fig 29. The lower scale is R.A., which for observers in the northern hemisphere increases from right to left; the upper scale is SHA, increasing by 15° per hour of R.A. in the opposite direction. The vertical scale on the right is declination, 15° being made the same length as 1 hour of R.A. The horizontal zero line is the celestial equator, the vertical one the hour circle of Aries, and their intersection the First Point, ♈. Most star maps mark the ecliptic (the broken line through ♈) but not latitude and longitude; they are included here for illustrative purposes and for use in one of the exercises. Junior drawing exercises can well be done on squared paper, using the scales suggested in Exercise 93 for small areas of sky and half that for larger ones. The axes should be drawn as in Fig 29, but of course for the particular range required—not just a copy. Seniors would do better to work on a larger scale on plain paper, ruling their own graticule (pattern of squares) at any convenient scale. In this case it is a great help to draw Fig 29A on thin card, to the same scale as your chart, and cut it out. Place it on your paper so that dec. 0° is on the equator; set the required minute of R.A. against the hour line of R.A. one greater than you want (eg for 2h 40m set 40m on the card against III on the chart); put your pencil against the required declination, and that is the spot. If you use a pencil in the left hand, draw the degrees on the left side of the card and set the minute of R.A. against the hour actually required.

For high latitudes polar co-ordinates must be used, Fig 30; a polar map is normally circular and the diagram is just an illustrative slice. Polar graph paper is expensive, so most users will have

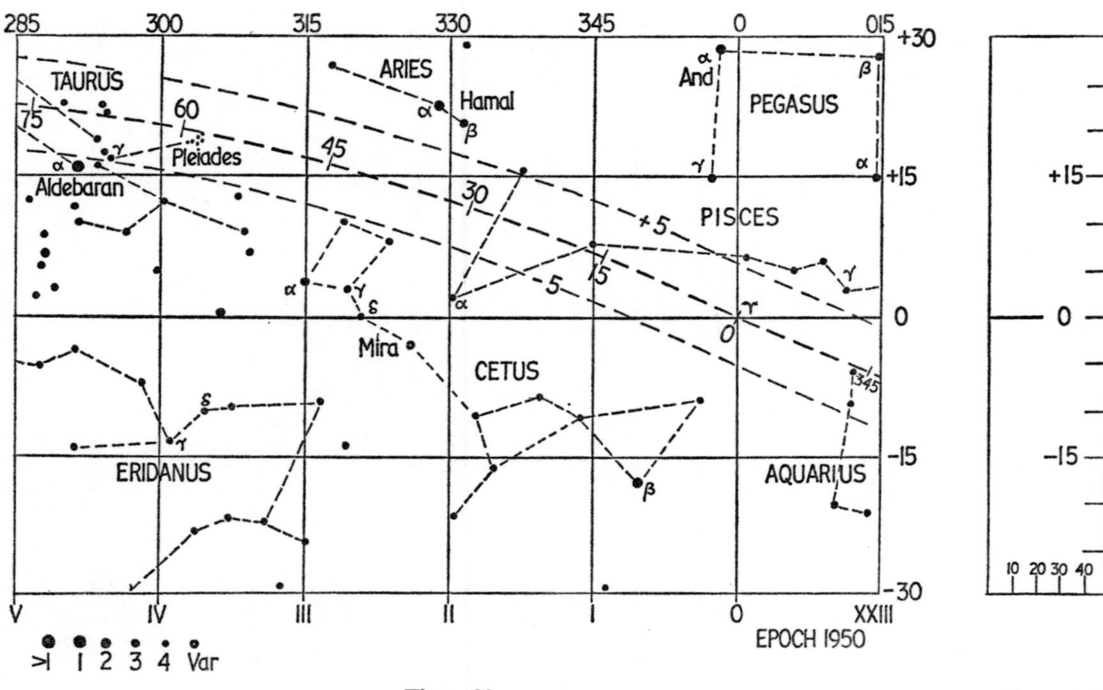

Figure 29

Figure 29A

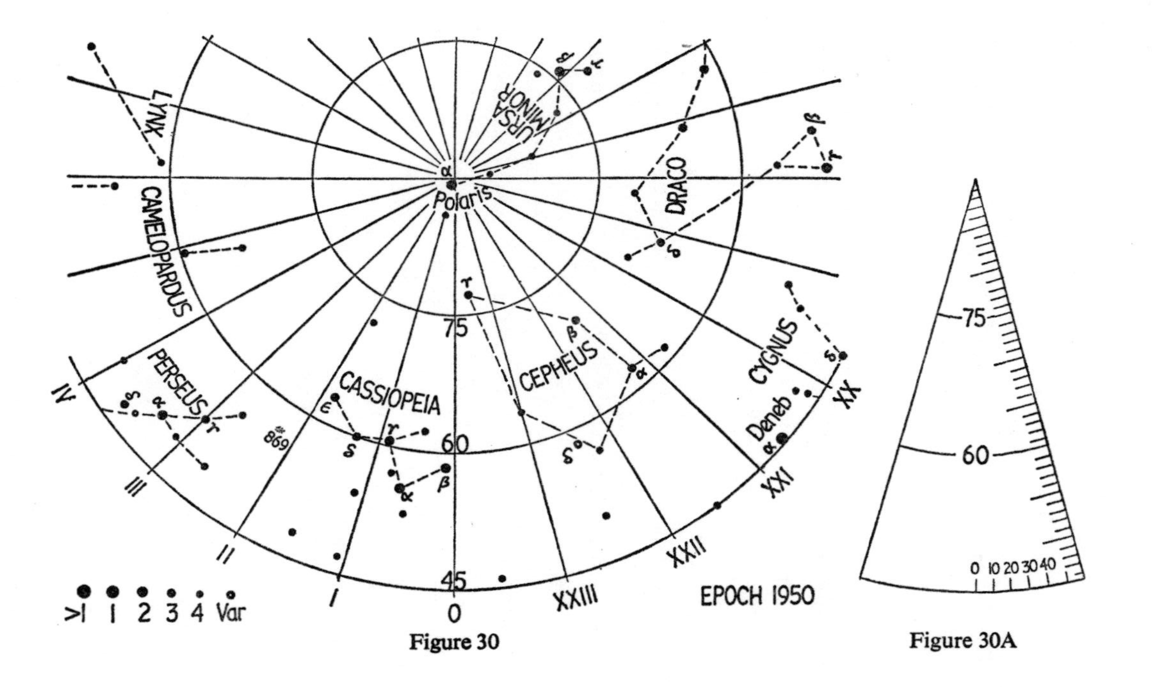

Figure 30

Figure 30A

to draw their own. (Schools possessing a spirit duplicator can prepare a stock.) Lines of R.A. radiate from the pole, 15° per hour apart, and are numbered clockwise for the northern hemisphere and anti-clockwise for the southern. The concentric circles are parallels of declination, and can be on any convenient scale. Fig 30A shows the card for measuring intermediate positions; for a southern map the minute scale should be numbered in the opposite direction and set to the hour actually required.

A star chart must indicate the magnitudes of the stars as well as their positions, and the method of using dots of different sizes is the commonest. As an aid to uniformity, make a stencil by drilling holes of different sizes in a plastic set-square; when using it your pencil must be needle-sharp or your star images will be neither uniform nor round. Include a key of magnitudes on the chart.

It has already been pointed out (page 41) that the pattern of R.A. and dec. is slowly changing owing to precession, and that a good star atlas bears its date or 'epoch'. Suppose, for instance, that an observer receives the position of a comet in the co-ordinates of 1970 and he wishes to find its position among the stars. His star atlas is Epoch 1920. He must therefore convert the 1970 comet to match the 1920 stars, and this he does by calculation or by the use of tables and graphs. Calculation in Smart; also, together with graphs etc, in J. B. Sidgwick, *Amateur Astronomer's Handbook*, London 1955. A simple table has been included in the 1971 edition of the annual *Handbook* of the BAA. As the correction amounts to only about 1° in a lifetime it is outside the 'precision limits' of this book. The stars themselves are not quite fixed on the celestial sphere, but changes due to this individual *proper motion* are usually smaller than the general precessional one.

STELLAR MAGNITUDE

The stars were classified in magnitudes many centuries ago. The twenty or so brightest stars were called 1st magnitude, meaning 1st class; then came about fifty of mag. 2, 2nd class, and so on. Thus the brightest stars have the lowest magnitude number. The ratio in apparent luminosity is about $2\frac{1}{2}$, so if the Pole Star is mag. 2, mag. 1 is about $2\frac{1}{2}$ times as bright and mag. 3 about $1/2\frac{1}{2}$. If two stars differ by two magnitudes their luminosities differ by $2{\cdot}5 \times 2{\cdot}5 = 6{\cdot}25$; 3 mag., $2{\cdot}5^3 = 15{\cdot}6$; 4 mag., $2{\cdot}5^4 = 39$; 5 mag., $2{\cdot}5^5 = 98$, or just about 100. Thus the faint naked-eye stars of the 5th magnitude (some people say that the 6th can be reached with the naked eye, but that calls for good sight and a very clear sky) are just about 16 times fainter than the Pole Star. These figures can be used for making rough estimates.

Example 1: What is the magnitude of a star 20 times brighter than the Pole Star? Three magnitudes gives a factor of 15·6, so in this case the difference is rather more than 3; call it 3·3. Then subtracting 3·3 from 2 leaves $-1{\cdot}3$; very bright stars have negative magnitudes.

Example 2: How much fainter than the Pole Star is magnitude 9 (just about visible with binoculars)?

The difference in magnitude is 7. A difference of 5 means a factor of 100, and a further difference of 2 gives another 6·25. Thus the star is fainter by $6{\cdot}25 \times 100 = 625$.

Needless to say, stellar magnitudes do not fall into precise groups, like the sizes of ball bearings. A nearer analogy is that of pebbles which have passed one riddle, but not the next; the selection includes every possible size between the two limits of wire mesh. It is therefore necessary to use a decimal subdivision of the magnitudes.

Now let us put this on a definite mathematical basis. A difference of 5 magnitudes is *by definition* 100. Then the light ratio between one magnitude and the next will be the fifth root of 100, which is 2·512 instead of the $2\frac{1}{2}$ of the last paragraph. Let l_1 and l_2 be the apparent luminosities of two stars and m_1 and m_2 their magnitudes. If l_1 is greater than l_2, m_2 will be greater than m_1, for the brighter the star the smaller its magnitude number. Then the ratio

$$\frac{l_1}{l_2} = 2{\cdot}512^{(m_2-m_1)}.$$

To solve this we must take logarithms:

$$\log\left(\frac{l_1}{l_2}\right) = (m_2 - m_1)\log 2{\cdot}512$$
$$= (m_2 - m_1) \times 0{\cdot}4.$$

Now apply it to the two examples previously done, giving the Pole Star a more correct value of 2·1 ($= m_2$, the fainter star, in example 1 and m_1 in 2).

(1)
$$\log 20 = (2{\cdot}1 - m_1) \times 0{\cdot}4$$
$$1{\cdot}301 = (2{\cdot}1 - m_1) \times 0{\cdot}4$$
$$2{\cdot}1 - m_1 = \frac{1{\cdot}301}{0{\cdot}4} = 3{\cdot}25$$
$$m_1 = -1{\cdot}15$$

(2)
$$\log\left(\frac{l_1}{l_2}\right) = (9 - 2{\cdot}1) \times 0{\cdot}4$$
$$= 2{\cdot}76$$
$$\frac{l_1}{l_2} = \text{antilog}\, 2{\cdot}76 = 575.$$

In the ratio l_1/l_2 always put the brighter on the top line, or there will be arithmetical troubles with the logarithms.

A double star is two in the telescope but one to the eye; let us consider how their magnitudes combine. The star Mizar has components of mag. 2·40 and 3·96. In order to keep the ratio l_1/l_2 greater than 1 we will compare them with a star of mag. 4, an imaginary one fainter than either and which we will call l_s.

$$\log\left(\frac{l_1}{l_s}\right) = (4 - 2{\cdot}4) \times 0{\cdot}4 \qquad \log\left(\frac{l_2}{l_s}\right) = (4 - 3{\cdot}96) \times 0{\cdot}4$$
$$l_1 = 4{\cdot}365 \times l_s \qquad\qquad l_2 = 1{\cdot}038 \times l_s$$

The total brightness $l = (4{\cdot}365 + 1{\cdot}038) \times l_s$. Then

$$\log\left(\frac{l}{l_s}\right) = \log 5{\cdot}403 = 0{\cdot}7326 = (4 - m) \times 0{\cdot}4$$

$$1{\cdot}83 = 4 - m$$

$$m = 2{\cdot}17$$

STELLAR PARALLAX

The parallax of the Moon was defined as the angle subtended at the Moon by the radius of the Earth (page 47). The parallax of a star is the angle subtended at the star by the radius of Earth's orbit. As the Earth moves from E_1 to E_2, Fig 31, the position of the near star S against the background of more distant stars moves from P_1 to P_2. The displacement, from which the angle of parallax π can be deduced, is very small, but can be measured down to about $0''{\cdot}01$. Parallax measured in this way is called *trigonometrical parallax*, to distinguish it from indirect methods of which you may have read elsewhere.

The astronomers' unit of distance is based directly on the parallax: if the parallax is 1 second of arc, the distance is 1 *parsec* (pc). So far no star as near as this has been found; the nearest star, Proxima Centauri, has a parallax of $0''{\cdot}763$ and a distance of $1/0{\cdot}763 = 1{\cdot}31$pc.

Now in radian measure

$$\pi = \frac{\text{arc}}{\text{radius}} = \frac{r}{d}; \quad d = \frac{r}{\pi} = \frac{1}{\pi}\text{A.U.}$$

But,

$$1'' \text{ of arc} = \frac{3{\cdot}14}{180 \times 3\,600}\text{rad. (page 4)}$$

so

$$1\text{pc} = \frac{180 \times 3\,600}{3{\cdot}14} = 206\,300\text{A.U.}$$

In km it becomes $206\,300 \times 150 \times 10^6 = 3{\cdot}095 \times 10^{13}$.

The popular unit of distance is the *light year*, the distance light, having a velocity of $3 \times 10^8\text{m s}^{-1}$, will travel in a year. This is

$$3 \times 10^5 \times 3\,600 \times 24 \times 365 = 9{\cdot}46 \times 10^{12}\text{km}$$

$$1\text{pc therefore} = \frac{3{\cdot}095 \times 10^{13}}{9{\cdot}46 \times 10^{12}} = 3{\cdot}26 \text{ light years.}$$

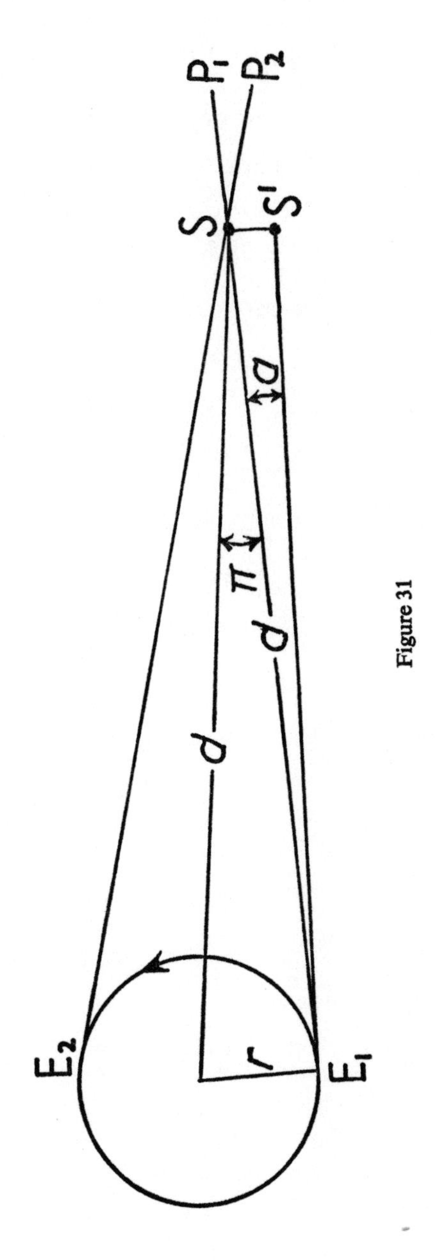

Figure 31

The above have been worked to our usual three-figure accuracy and therefore differ slightly from the 'table book values', which are as follows:

$$1 \text{ light year} = 9{\cdot}4607 \times 10^{12}\text{km}$$

1 parsec = $3{\cdot}0857 \times 10^{13}$km or 206 300A.U. or 3·2616 light years.

A note should be added that the symbol π is used for stellar parallax, as well as for the more usual circumference/diameter ratio. Both have turned up in this section, but the latter is represented only in the numerical form 3·14.

ABSOLUTE MAGNITUDE

So far we have compared the luminosities of stars as we see them, but if we find that one star is twice as bright as another it does not follow that it is really twice as bright, for it might be nearer. To get a ratio of true luminosities the astronomer calculates what the magnitudes would be if the stars concerned were both at a distance of 10 parsecs, and then uses those to find the ratio. The *absolute magnitude* of a star is the magnitude which it would have if its distance were 10 parsecs. Let L and M be the luminosity and magnitude at this distance, and l and m the values at its actual distance of d pc. If L is greater than l, then M will be less than m, so we can write

$$\frac{L}{l} = 2{\cdot}512^{(m-M)}.$$

Now the intensity of light varies inversely as the square of the distance between the source and the observer, so we can write

$$\frac{L}{l} = \frac{d^2}{10^2}.$$

Hence

$$2{\cdot}512^{(m-M)} = \frac{d^2}{10^2}$$

$$(m - M)\log 2{\cdot}512 = 2\log d - 2\log 10$$

$$(m - M) \times 0{\cdot}4 = 2\log d - 2$$

$$-M = 5\log d - 5 - m$$

or

$$M = m + 5 - 5\log d$$

Example: The apparently brightest star is Sirius, mag. $-1{\cdot}44$, distance 2·7pc. Compare its true luminosity with that of Spica, mag. 0·97, dist. 65pc.

First find their absolute magnitudes:

Sirius	Spica
$M = -1{\cdot}44 + 5 - 5\log 2{\cdot}7$	$M = 0{\cdot}97 + 5 - 5\log 65$
$= 3{\cdot}56 - 5 \times 0{\cdot}4314$	$= 5{\cdot}97 - 5 \times 1{\cdot}813$
$= 3{\cdot}56 - 2{\cdot}16 = +1{\cdot}40$	$= 5{\cdot}97 - 9{\cdot}06 = -3{\cdot}09$

Then

$$\frac{\text{Spica}}{\text{Sirius}} = 2{\cdot}512^{(1{\cdot}4+3{\cdot}09)}$$

$$\log\left(\frac{\text{Sp}}{\text{Si}}\right) = 4{\cdot}49 \times 0{\cdot}4 = 1{\cdot}80 \quad \frac{\text{Sp}}{\text{Si}} = \text{antilog}\,1{\cdot}80 = 63.$$

So Spica is 63 times as luminous as Sirius. It is nearly double the diameter and has twice the surface temperature.

The astronomer has means of inferring absolute magnitude from physical characteristics such as spectrum and variability, and he can use this for finding distance—particularly useful when it is too great for the trigonometrical method.

Example: Find the distance of Regulus, given that its apparent and absolute magnitudes are 1·34 and –0·8 respectively.

$$-0{\cdot}8 = 1{\cdot}34 + 5 - 5\log d$$

$$5\log d = 7{\cdot}14$$

$$\log d = 1{\cdot}43$$

$$d = \text{antilog}\,1{\cdot}43 = 27\text{pc}$$

There are other ways of reckoning magnitudes besides the visual. For example, photographic comparison gives different results from visual, because the response of a photographic film to different colours is not the same as that of the eye. This, however, does not affect the foregoing work provided that you know which system you are using—and keep to it.

THE MASS OF A BINARY STAR

Double stars have already turned up. When the two components are near enough to each other to be observed in motion, the term *binary star* is used. They are, of course, both in motion about the common centre of gravity, like the Earth–Moon system, but also like the Earth–Moon system it is quite valid to regard the lesser star to be in orbit about the greater. The period and the angular separation can be observed, though the latter will not be the true separation a because the orbit is likely to be inclined to the line of sight and what we see is the projection of it onto the celestial sphere. However, when the pair has been observed for a period of years the true value of a can be found. Turn back to Fig 31. S and S′ are the two components, a the true angular separation and π the parallax. We have already seen that $d = 1/\pi$ in A.U.; it is also equal to SS′/a. Thus

$$\frac{1}{\pi} = \frac{\mathrm{SS}'}{a}; \quad \mathrm{SS}' = \frac{a}{\pi} \text{ in A.U.}$$

The last chapter included the universal form of Kepler's third law,

$$M + M' = \frac{D^3}{T^2},$$

when M is in sun-masses, D in A.U. and T in years. For the binary star SS′ is D, the observed period is T, so the mass of the system can be found.

Example: The binary system Rigil Kent (α Centauri) has a parallax of 0″·732, a true separation of 17″·7 and a period of 80·1 years.

$$\mathrm{SS}' = \frac{a}{\pi} = \frac{17{\cdot}7}{0{\cdot}73} = 24{\cdot}2 \text{A.U.},$$

$$M + M' = \frac{(24{\cdot}2)^3}{(80{\cdot}1)^2} = 2{\cdot}2 \text{ sun-masses.}$$

The 'table book value' for this star is almost exactly 2, but 2·2 happens to be the average value for binary stars in general.

This calculation, in reverse, is another of the indirect methods of finding stellar distances. By assuming that the pair is an average specimen with $M + M' = 2{\cdot}2$ and combining it with the observed period T, the separation SS′ in A.U. can be calculated. Comparing this with the observed true angular separation a we can find the parallax from $\mathrm{SS}' = a/\pi$, or the distance in A.U. from $d = \mathrm{SS}'/a$. Trying this out will be left to the student (Ex. 142).

THE DOPPLER EFFECT

It is assumed here that previous reading has provided some knowledge of the nature of electromagnetic waves and of the spectrum. If not, this section will be unintelligible and should be omitted. The Doppler effect is a well-known phenomenon in sound: the change in pitch of the note heard from a low-flying aircraft as it passes overhead. As it approaches, the apparent wavelength is shorter and the pitch higher than their real values, and the opposite when receding. The same applies to light and radio waves from celestial objects. A velocity of approach gives a shorter wavelength, so spectral lines are displaced across the band of colour towards the blue end; a recession gives a lengthening of the waves and a displacement towards the red. The colours themselves do not move and we can regard them as created in the eyes and brain of the observer.

Let the source emit f waves per second (the frequency). If each wave has a length of λ, at the end of one second the leading wave will have travelled $f \times \lambda$; this is the velocity, which we will call c. Therefore

$$c = f\lambda \quad \text{or} \quad \lambda = \frac{c}{f}.$$

Now suppose that the source is moving towards the observer with a velocity v; the tail of the wave train is chasing its head. This means that the f waves have been compressed into a length $c - v$, so the apparent wavelength seen by the observer is $(c - v)/f$. The difference

$$\lambda - \lambda' = \frac{c}{f} - \frac{c - v}{f} = \frac{c}{f} - \frac{c}{f} + \frac{v}{f} = \frac{v}{f}.$$

But

$$\lambda = \frac{c}{f}.$$

Dividing, we get

$$\frac{\lambda - \lambda'}{\lambda} = \frac{v}{f} \div \frac{c}{f} = \frac{v}{c}.$$

The difference $\lambda - \lambda'$ is often expressed as $\delta\lambda$ so the final formula becomes

$$\frac{\delta\lambda}{\lambda} = \frac{v}{c}.$$

If the source were receding the train of waves would be stretched out giving an apparent wavelength $(c+v)/f$, λ' now being greater than λ. Then

$$\lambda' - \lambda = \frac{c+v}{f} - \frac{c}{f} = \frac{v}{f}$$

and we should get the same final formula as before (see also Note 9, page 106).

The velocity of electromagnetic radiation is 3×10^8m s^{-1}. Radio wavelengths are normally expressed in metres, so direct division gives frequency in cycles per second, a unit now called the hertz.

Example: To find the frequency of 3m waves.

$$f = \frac{c}{\lambda} = \frac{3 \times 10^8}{3} = 10^8\text{Hz}.$$

Wavelengths in visible light, ultra violet etc., are very short in comparison and are usually expressed in ångström units, one such unit being 10^{-10}m. The line known as Hα has a wavelength of 6 562Å; before finding the frequency this must be changed to $6\,562 \times 10^{-10}$m, and as this is 0·0000006562m the words 'very short' were quite justified. Conversion to metres is not necessary when using the formula for Doppler shift, as the wavelengths are used in ratio and units have disappeared.

Example: What would be the displacement of the Hα line if the source were approaching with a velocity of 1000km s^{-1} (10^6m s^{-1})?

$$\frac{\delta\lambda}{6\,562} = \frac{10^6}{3 \times 10^8} \quad \delta\lambda = \frac{10^6 \times 6\,562}{3 \times 10^8} = 21{\cdot}8\text{Å to blue.}$$

Symbols and Abbreviations

♈	First Point of Aries
♎	First Point of Libra
☊	Ascending node, longitude of
☋	Descending node, longitude of
° ′ ″	Degrees, minutes and seconds of arc
α (alpha)	Right ascension
β (beta)	Celestial latitude
δ (delta)	Declination
Δ (delta)	Distance of comet from Earth
ϵ (epsilon)	Obliquity of the ecliptic
θ (theta)	Terrestrial longitude, angles in general
λ (lambda)	Celestial longitude, wavelength
π (pi)	circum./diam. = 3·1416, stellar parallax
ϖ ('curly' pi)	Longitude of perihelion (planet)
ϕ (phi)	Terrestrial latitude
ω (omega)	Argument of perihelion (comet)
a	Semi-major axis of ellipse, mean distance
Å	Ångstrom unit
A.U.	Astronomical unit
b	Semi-minor axis of ellipse
c	Velocity of light
cgs	Centimetre-gramme-second system of units
e	Eccentricity
E	Equation of time
f	Frequency
g	Acceleration due to gravity
G	Gravitational constant
GCT	Greenwich civil time
GHA	Greenwich hour angle
GMAT	Greenwich mean astronomical time
GMT	Greenwich mean time
h m s	Hours, minutes and seconds

H.A.	Hour angle
H.A.♈	Hour angle of Aries
HAMS	Hour angle of mean Sun
HATS	Hour angle of true Sun
H.P.	Horizontal parallax
Hz	Hertz, or cycles per second
i	Inclination to the ecliptic
l.y.	Light year
LHA	Local hour angle
LMT	Local mean time
MT	Mean time
n	Mean daily motion of planet
pc	Parsec
P	Period of revolution
q	Perihelion distance
r	Distance of comet from Sun
R.A.	Right ascension
SHA	Sidereal hour angle
SL	International units (Système Internationale d'Unités)
ST	Sidereal time
UT	Universal time

Almanacs

The Astronomical Ephemeris is the standard work for astronomers, but is unnecessarily large and expensive for likely users of this book. *The Nautical Almanac* is specially arranged for use in navigation, and some of the material is presented in a form not very convenient for the ordinary astronomer; if it happened to be available it could be used. *Whitaker's Almanack* is a general reference annual, but with some eighty pages or so of astronomical information it covers most of the amateur's needs as far as numerical astronomy is concerned. *The Handbook of the British Astronomical Association* is prepared specially for observers of the various heavenly bodies; it contains some of the information which is in Whitaker and much which is not. The contents of these two, so far as it applies to the present book, is listed below. The *Handbook* can be obtained from the British Astronomical Association, Burlington House, London, W.1, and the other three from booksellers.

The Astronomical Ephemeris and *The American Ephemeris* is a joint British–American production, and is therefore available on both sides of the Atlantic. In North America the equivalent of the BAA *Handbook* is *The Observer's Handbook*, published by and obtainable from The Royal Astronomical Society of Canada, 252 College Street, Toronto 2B, Ontario. The content of the Canadian handbook is not quite the same as that of the British one but its purpose is to serve the practical amateur observer. Several annual books supplying numerical data are also published in the United States and readers in that country should seek advice from their local astronomical society.

HANDBOOK	WHITAKER	INFORMATION
Equation of time	Daily to 1s	—
Sidereal time at 0h UT	Daily to 1s	4-day intervals to 0·1m
Mean time transit of ♈	Daily to 1s	—

INFORMATION	WHITAKER	HANDBOOK
Sun:		
R.A. and dec.	Daily	4-day intervals
longitude	—	4-day intervals
semi-diameter	—	4-day intervals
transit of true sun	Daily to 1m	4-day intervals to 0·1m
rising and setting	Daily	5-day intervals
Moon:		
R.A. and dec.	Daily	—
rising and setting	Daily	—
phases	Included	Included
perigee and apogee	Included	Included
H.P. and semi-diameter	Daily	Twice a lunation
longitude of ☊	1st of each month	—
Eclipses during the year	Included	Included
Planets:		
time of transit	Every 3–10 days	—
R.A. and dec.	Every 3–10 days	Every 5–10 days while the planet is observable
distance, magnitude	—	(Every 5–10 days while the planet is observable)
elements of orbits	—	Included
Comets expected during the year	—	Elements and ephemerides
Co-ordinates of the principal stars	Included	—

Notes

Note 1

An accurate method of drawing an ellipse is shown in Fig 32. Draw the axis and put in C and F, remembering that CA $= a$ and CF $= ae$. Find b either geometrically as on page 12 or from $b = a\sqrt{1-e^2}$. Draw the right-angled triangle so that LM $= a$ and MN $= b$. (Squared paper is useful here.) With centre C and radius a describe a circle. Draw a number of perpendiculars to the axis cutting the circle, YKY′ being one such line. On LM make LV = KY; draw the perpendicular VW; make KP = KP′ = VW. Then P and P′ are two points on the ellipse. Plot other pairs of points in the same way.

Note 2

The equation for the ellipse is

$$\frac{x^2}{a^2} + \frac{y^2}{b^2} = 1$$

whence the ordinate

$$y = \frac{b}{a}\sqrt{a^2 - x^2} \quad (= \text{KP when } x = \text{CK}).$$

For a circle

$$x^2 + y^2 = a^2$$

whence

$$y = \sqrt{a^2 - x^2} \quad (= \text{KY when } x = \text{CK}).$$

Thus if P is on the ellipse the ratio KP/KY would be equal to b/a. It is, because we have made

$$\frac{\text{KP}}{\text{KY}} = \frac{\text{VW}}{\text{LV}} = \frac{b}{a}.$$

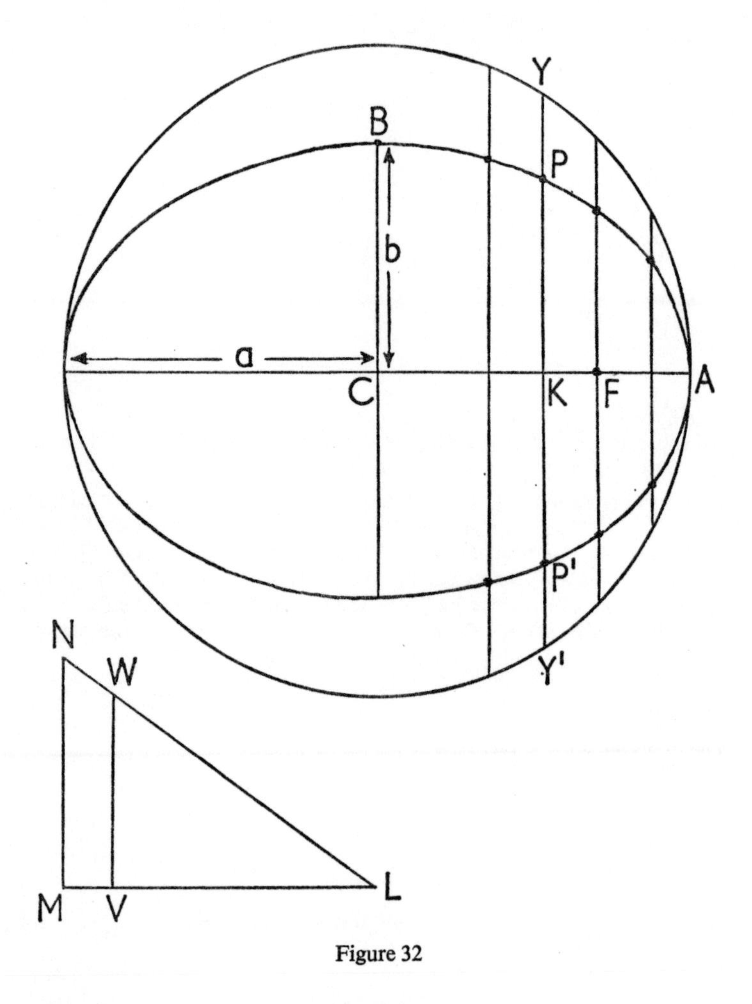

Figure 32

Note 3

In co-ordinate geometry the equation for the ellipse is

$$\frac{x^2}{a^2}+\frac{y^2}{b^2}=1$$

when the centre C is taken as origin. In Fig 6 the co-ordinates of P are $x = a\cos\theta$ horizontally and $y = b\sin\theta$ vertically. Substituting in the equation we get

$$\frac{a^2 \cos^2\theta}{a^2} + \frac{b^2 \sin^2\theta}{b^2} = \cos^2\theta + \sin^2\theta = 1.$$

Thus P satisfies the equation and does lie on the ellipse.

Note 4

The inclination is variable, the angular sizes of the Sun and Moon are not constant, and their motions in their orbits are not uniform. (See Barlow and Bryan.)

Note 5

Users of old records should know that until 1925 Jan 1, GMT was reckoned from noon, and was therefore simply HAMS. This reckoning was then renamed Greenwich Mean Astronomical Time (GMAT), and for some years what we now know as GMT or UT was called Greenwich Civil Time (GCT).

Note 6

The Gregorian calendar was adopted in England in 1752, when there was an accumulated error of 11 days. To correct this, Sept 2 was followed by Sept 14. Anyone calculating a time interval which overlaps this period must remember not to include 'the lost eleven days'. Another trap for the historian is the fact that 1752 began on Jan 1, whereas under the then existing calendar, 1751 was not due to end until March 25. Three months belonged to both years and must not be counted twice.

Note 7

Teachers embarking on this chapter will find it helpful to have a chalk-board globe, as some pupils have difficulty in changing a two-dimensional diagram into a three-dimensional conception. The construction of a permanent celestial sphere is described by Tricker. For juniors, the 'Tancock flask' (Fig 33) is helpful. By ignoring the glass on the near side, they can imagine that they are inside the sphere and watch rising and setting, circumpolar stars, and other aspects of this chapter.

Note 8

The angle from the perihelion to the point obtained (the angle ϖSM) is the mean *anomaly*, and the true anomaly is found by adding a quantity called the *equation of centre*. In terms of the 'precision limits' of this book the correction is small, though it can rise to 10° for Mars and over 20° for Mercury. Readers with a little trigonometry may like to estimate the equation of centre in degrees from $(360\epsilon \sin M)/\pi$, where M is the mean anomaly. Stetson gives tables for this correction.

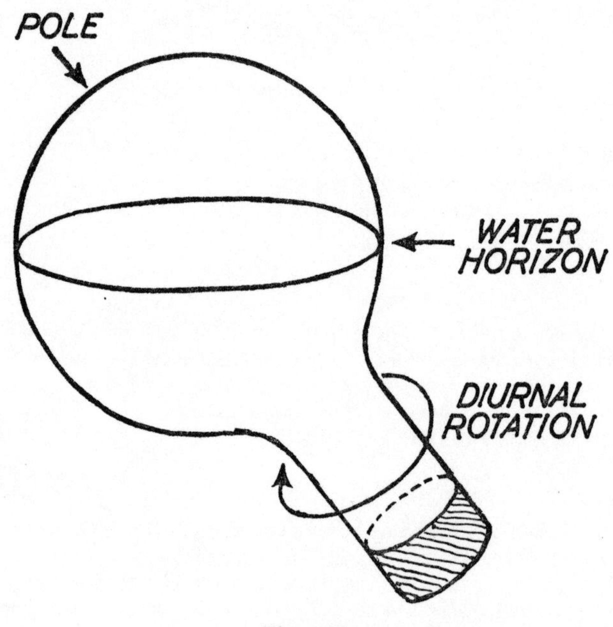

Figure 33

Note 9

This formula is valid only if $\delta\lambda/\lambda$, commonly represented by z, does not exceed about 0·15, which corresponds with a velocity of the source nearly one-sixth that of light. For higher velocities than this the principle of relativity demands a more complex relationship.

References

Astronomical Epheremis, annual, see page 101.
Barlow, C. W. C. and Bryan, G. C. *Mathematical Astronomy*, 5th ed, revised by Spencer Jones, Sir Harold, 1944. A useful reference book for amateurs and students, though more advanced than the present one.
Běcvář, A. *Atlas Coeli Katalog 1950*, Prague, 1960. Used, with some simplification, as the principal source book for Chapter 5.
Burns, P. F. *First Steps in Astronomy*, 1942. Further ideas and exercises for junior students in schools.
Davidson, M. *Elements of Mathematical Astronomy*, 3rd ed, revised by Dinwoodie, C. 1962. Written for advanced amateur astronomers.
Handbook of the British Astronomical Association, annual, see page 101.
Minneart, M. G. T. *Practical Work in Elementary Astronomy*, Dordrecht, Netherlands, 1969. An undergraduate university course.
Nautical Almanac, annual, see page 101.
Smart, W. M. *Textbook of Spherical Astronomy*, 5th ed, Cambridge, 1962. A university textbook.
Stetson, H. T. *Manual of Laboratory Astronomy*, Boston, 1928. A university course with a practical emphasis.
Tricker, R. A. R. *The Paths of the Planets*, 1967. For sixth forms and amateur astronomers. See page 59.
Whitaker's Almanack, annual, see page 101.

Acknowledgements

The author is greatly indebted to the authors, compilers and publishers of those books listed in the References, for all these have been valuable source material for the present book. Special thanks are extended to the Superintendent of the Nautical Almanac Office and to the Council of the British Astronomical Association.

For encouragement and advice in the planning and reading of the manuscript, Dr J. G. Porter and Dr V. Barocas have been most helpful at all times, while Patrick Moore has done a considerable amount to help in the publication of this book.

Exercises

EXERCISES ON CHAPTER 1

Group 1 : Practical

1 To find the meridian from first principles.

Mount a vertical rod AB (Fig 34) on a smooth horizontal surface, 30–50cm high if you are working on a table and about 150cm out in the yard. It is common knowledge that the Sun is highest, and hence the shadow cast by the rod shortest, when crossing the meridian. From this it follows that equal shadows will be formed at equal times before and after the shortest. Sometime before noon mark the direction of the shadow BC and measure its length. Then *either* repeat every few minutes until you find the shadow lengthening again, when the shortest recorded position will mark the meridian, *or* leave it until sometime after noon, and then test it until you find the length BD the same as the first measurement. Bisect the angle between this shadow and the first one. A string tied at B and knotted at C gives a simple way of finding when BD is the same length as BC.

2 To check the daily motion of the Moon.

Make two cardboard screens, each with a vertical slot in it. Place them so that the two slots are over the meridian and about 30cm apart. Time as near as you can when the west limb of the Moon can be seen through both slots, and remember to check your watch with a time signal. Repeat for several evenings and compare with the remarks on page 17. This is easiest to do in the summer evenings when the Moon is low in the sky.

3 To find your latitude.

Make a simple clinometer, Fig 35A, unless a proper one is available. Look at the Pole Star through the slots, taking care that the plumb-line hangs free. When all is steady, nip the thread

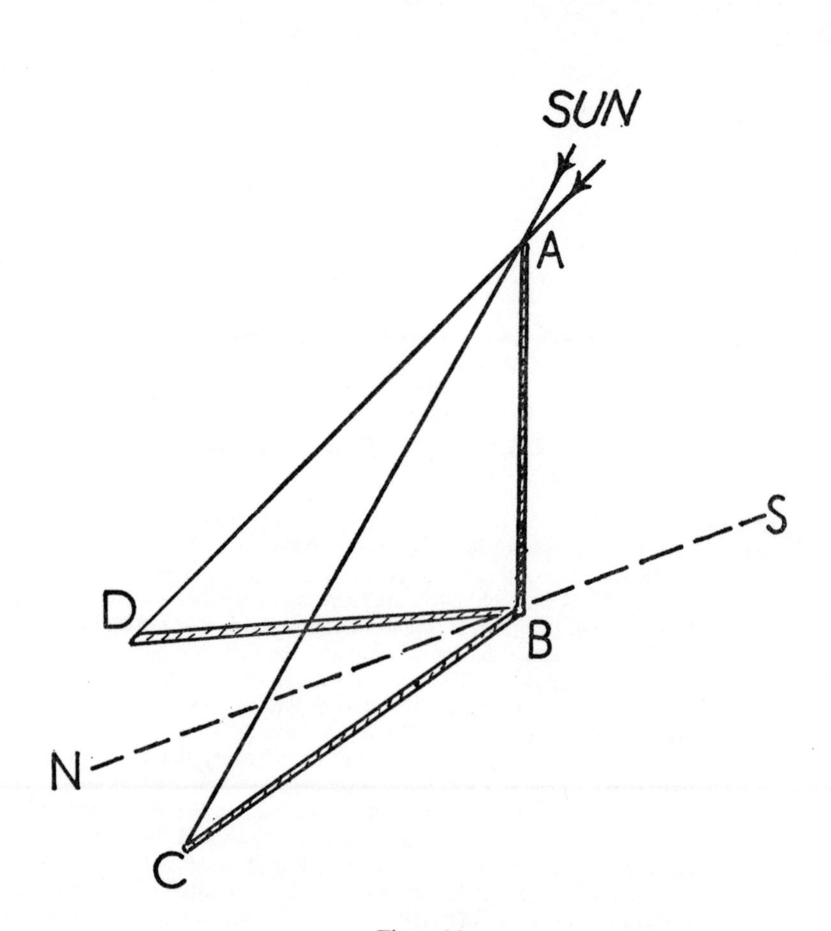

Figure 34

against the protractor with your finger and thumb, take it to the light and read the angle. The altitude of the pole is equal to the latitude of the place, Fig 35B. The observer is at O and OH is his horizon; angle OCE is his latitude. As the pole is very distant CP and OP′ are parallel, so the corresponding angles ZOP′ (the *zenith distance*) and OCP (the *co-latitude*) are equal. Therefore their complements, altitude HOP′ and latitude OCE, are equal. The Pole Star is not quite at the pole, but just under a degree from it, so your result would not ordinarily be exact even if your observation were perfect.

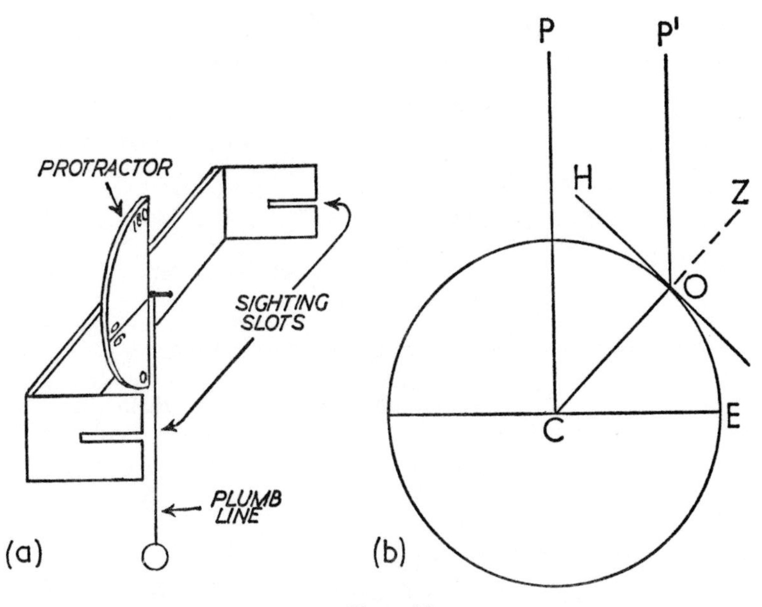

Figure 35

Group 2: Recalling Previous Reading

4 Explain how the inclination of the Earth's axis causes the seasonal changes in our climate.

5 Draw a diagram illustrating the phases of the Moon. Indicate in particular (i) first quarter, (ii) a waning crescent, (iii) waxing gibbous. At what time of day would you look for each of these three?

6 Show by diagrams of the shadows the formation of (i) a solar eclipse, (ii) a lunar eclipse. Distinguish between partial and total eclipses in each case.

Group 3: General

7 The *ellipticity* of a planet is defined as the difference between the equatorial and polar radii expressed as a fraction of the equatorial. Calculate the ellipticity from the figures on page 3.

8 Use an atlas to find the latitude and longitude of (i) Edinburgh, (ii) Montreal, (iii) Los Angeles, (iv) Sydney, (v) Moscow, (vi) Johannesburg.

9 Convert into degrees and minutes (i) 0·3 radian, (ii) 1·1 radian; convert to radians (iii) 30°, (iv) 47°28′.

10 Find the angular diameter in minutes of arc of an object 10cm in diameter when viewed from a distance of 20m.

11 Write a short account of the two principal motions of the Earth and the relationship between them. What do you understand by 'a frame of reference'? Have you used one in this question?

12 Draw half ellipses (BAB′) by method 1, given that (i) $a = 10$cm, $e = 0{\cdot}2$, (ii) $a = 10$cm, $e = 0{\cdot}7$.

13 Draw ellipses by method 2, given that (i) $a = 6$cm, $e = 0{\cdot}1$, (ii) $a = 6$cm, $e = 0{\cdot}9$.

14 Draw ellipses by method 3, given that (i) $a = 5$cm, $b = 3$cm, (ii) $a = 5$cm, $e = 0{\cdot}4$.

15 Draw the orbit of the Earth by the method described on page 13. Retain your drawing for further use. Deduce e and ϖ.

16 By the method of page 13 draw the orbit of the Moon and find e. Retain your drawing for future use. A = geocentric longitude, B = angular diameter.

Date		A	B	Date		A	B
1970				1970			
Nov	1	245°	31′·0	Nov	15	080°	31′·2
	3	272	31·5		17	106	30·4
	5	299	32·0		19	131	29·9
	7	327	32·3		21	154	29·6
	9	356	32·5		23	178	29·7
	11	025	32·4		25	202	30·1
	13	053	31·9		27	228	30·8

The Moon crossed the ecliptic from S to N on the 7th and from N to S on the 20th. Insert the approximate positions of the nodes ☊ and ☋.

17 From the table on page 13 find the dates nearest to (i) aphelion, (ii) perihelion, and, from the table in Ex. 16, (iii) perigee, (iv) apogee.

18 With reference to the same tables, what would you see if the Moon passed centrally across the Sun on 1 Nov?

19 What is meant by the synodic and sidereal periods of the Moon and why are they different?

20 Draw the phases of the Moon (i) 5 days and (ii) 17 days after new moon.

21 Make an accurate drawing of an evening crescent moon when the illuminated part is one-quarter of the diameter (ellipse, method 3, diameter not less than 8cm).

22 Using Fig 2 and similar triangles calculate the length of the Earth's shadow, taking the diameters of the Earth and Sun to

be 12 700 and 1 390 000km respectively and the distance of the Sun to be 150×10^6km. (In questions 22 and 23 you can assume that the lengths of the shadows and the Earth–Moon distance are small enough to neglect when added to or subtracted from the distance of the Sun.)

23 (i) Find the length of the Moon's shadow, using data from Ex. 22 and taking the diameter of the Moon to be 3 500km. (ii) What is the diameter of the shadow at the nearest point on the Earth's surface if the eclipse takes place on a day when the Moon is 380 000km from the Earth, centre to centre? Radius of the Earth 6 350km.

24 Starting from the result of Ex. 22, (i) what is the diameter of the Earth's shadow at the distance of the Moon, 380 000km? If the period of the Moon is 27d 8h what is (ii) its velocity in km per hour? (iii) What is the maximum time for which the Moon (diam. 3 500km) can be completely immersed in the Earth's shadow?

25 In Fig 6 the new and full moons are not 180° apart. Why not?

26 What eclipses would you expect to take place if the new moon occurred (i) 15° before the node, (ii) 12° after the node?

EXERCISES ON CHAPTER 2

Group 1 : Practical

27 To find the meridian, using an almanac.

From the almanac find the time of transit of the true sun. Find your longitude from a map, the larger the scale the better; convert it to time units (page 26). If you are west of Greenwich the transit will be later, so add the longitude; if east, subtract. Check your watch against the BBC time signal or the telephone, so that you can deduce exact GMT (UT) from it. Over a smooth level surface hang a plumb-line from a stand of some sort. At the exact predicted time of solar noon mark the direction of the shadow of the line. It is helpful if this meridian could be marked somewhere permanently so that it is always available in future.

A school might build a brick pillar about 3 feet high, topped with a smooth and truly horizontal slab of slate or concrete about 2 feet square. It is useful as a telescope stand as well as for meridian experiments.

28 To find the meridian, using the map only.

Choose a distant object which you can locate on the Ordnance Map (1 inch or 2½ inch). Rule a line through the object and your point of observation. Place the map on a horizontal surface with the observing slits made for Ex. 2 standing on the line. Turn the map and the slits until you can see the object through them; the map is now 'set'. The grid lines will be very nearly parallel with the meridian, and somewhere in the margin there will be a note giving the difference, if any, between grid north and true north. If some readily identifiable object is actually on your meridian make a note of it; you can then use it to locate the meridian any time in daylight.

29 To find the meridian, using a compass.

The compass must be a good one, with a scale of degrees marked on a revolving card. From the Ordnance Map find the magnetic variation in your area, and correct it (as instructed on the map) for the lapse of time since it was printed. Place the compass on a horizontal surface and let it come to rest. If the variation is x°W, then the true north will be x°E of the arrow on the compass card, and vice versa. There must be no ironwork, such as an iron fence, within about 10 feet of your point of observation. This method is the least reliable of the series, as magnetic variation can differ locally from the area value given on the map. Magnetic variation is also called declination; if the term turns up anywhere be careful not to confuse it with astronomical declination, which is to be used in the next chapter.

30 To attempt to measure the equation of time.

Find your longitude from a map, convert to time, and calculate the GMT of 12h LMT (page 28). Check your watch so that you can read correct GMT from it. Place your pair of vertical slits (Ex. 2) on the meridian, and record the time at which the bright bar of light passing through the first slit falls exactly on the second. Do *not* look *at* the Sun. The difference between your observed time, the time of apparent noon, and 12h LMT is the equation of time. The observation is easiest to make in the winter when the Sun is low in the sky. This exercise is less satisfactory than it sounds; *don't* be disheartened if the result is poor, and *do* consider carefully and write down your difficulties and their possible causes. We shall return to this problem at a later stage.

31 To make a sundial.

Fig 36 illustrates the idea. ABCD is a horizontal plane and MM the meridian. OPQ is a rod pointing to the celestial pole and therefore in the plane of the meridian; it is called the gnomon. PCD is a plane perpendicular to the rod and therefore in the plane of the celestial equator (page 22). As the Sun moves across the sky in a direction parallel with the equator the shadow of the rod moves across PCD at the same rate of 15°

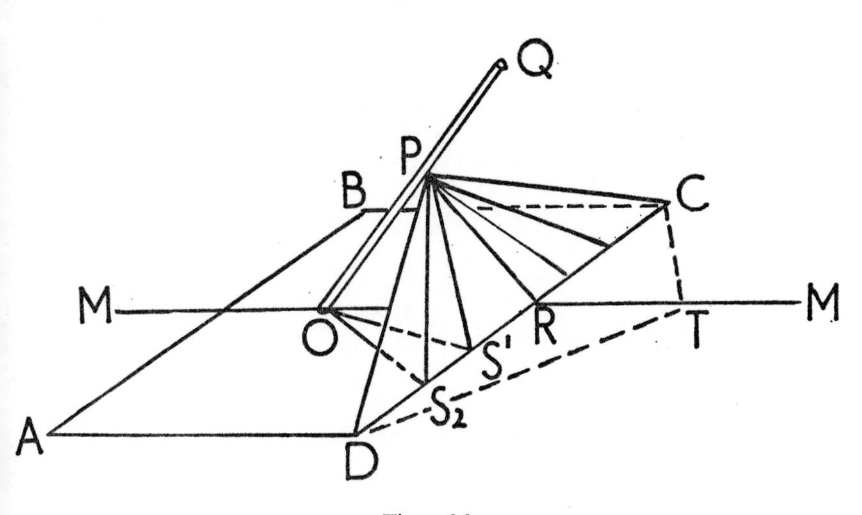

Figure 36

per hour. At noon the shadow falls on PR, an hour later on PS_1, another hour and it is on PS_2 and so on. $S_1, S_2 \ldots$ are the points at which the shadow strikes the horizontal plane ABCD and will enable the shadow lines to be drawn on that plane. PCD has now served its purpose. Imagine it to be rotated about CD until P falls on T, and that it is transparent; it will then make Fig 37, which we are about to draw.

Draw a triangle OPR so that OR is about the same length as the diameter of the dial you intend to make, angle ROP is equal to the latitude of the place at which you are going to use it, and angle OPR is 90°. Draw a horizontal line OT across a piece of good-quality paper which is very long from top to bottom—longer than suggested by Fig 37B. Make OR and RP the same length as they are in the triangle. At R draw a long straight line at right angles to OT. At P make the angles $RPS_1 = 15°$, $S_1PS_2 = 15°$, and so on, on both sides of PR. Join $S_1, S_2 \ldots$ to O. Draw a circle of the size of the intended dial, as shown, and where OR, $OS_1, \ldots$ cut the circle, ink in and number the hour marks. If you care to make 5° angles from P you could have 20-minute graduations and then subdivide those by eye. Stick the triangle OPR on to strong cardboard or thin wood and cut it out; the part to the right of the wavy line can now be discarded. This is the gnomon; mount it vertically on your baseboard (the 'carpentry' of the job is left to you). Cut out your graduated circle and divide it into halves along the line OR. Stick these to the baseboard, one on each side of the gnomon, and with O level with its foot. It is then ready for use. Place it on the meridian; make simultaneous readings of the dial and GMT once a day for at least a month. Plot the differences

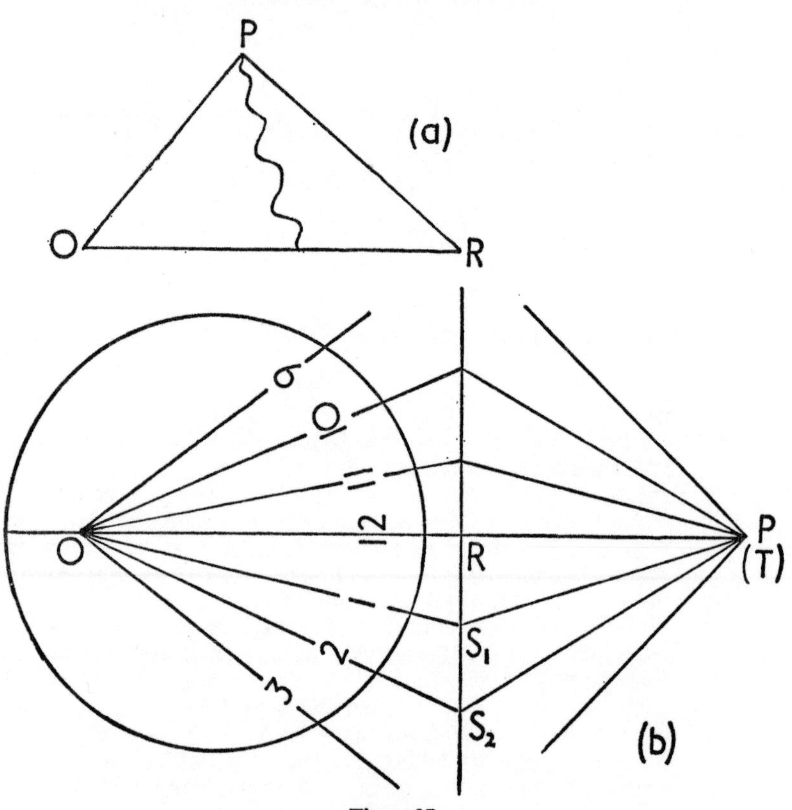

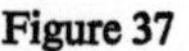
Figure 37

sundial–GMT on graph paper with differences as ordinate (vertical) and date as abscissa (horizontal). Explain the shape that you find. (As an alternative to drawing, or when RS becomes inconveniently long, the graduation marks can be calculated from the formula $\tan\theta = \sin \text{lat.} \times \tan 15n$, where n is the time interval in hours on either side of the meridian and θ the corresponding angle at O.)

32 To find your longitude.

Experimentally it is the same as Ex. 30. The difference between your GMT of transit and the Greenwich transit of the true sun from the almanac is your longitude in time. Convert into arc.

Group 2: Recalling Previous Reading

33 What is the relationship between Greenwich Mean Time, British Standard (Summer) Time, and the reading of the sundial in the churchyard?

34 There have been reports in the press about hazards to the health of the business man who makes frequent air journeys between Europe and America, and consequently suffers from abrupt changes in the timing of his daily routine. What changes, and why do they occur?

35 What changes in the appearance of the night sky would you expect to see between (i) 7 pm and 11 pm on the same evening, (ii) 8 pm on March 1 and 8 pm on September 1?

36 Why do we sometimes have a February 29, and what would be the consequence if we did not?

Group 3: General

37 Define sidereal time, solar time and mean time. How do they differ, and why?

38 What is meant by the equation of time? If its value is +8m, at what time will the Sun be observed to cross the Greenwich meridian?

39 Suppose that the day (sunrise to sunset) is exactly 10 hours long, and that the equation of time for that day is –6m. What will be the times of sunrise and sunset as recorded by (i) a sundial, (ii) a mean time clock? Is it right to say that mornings and afternoons are equal in length?

40 Draw a graph showing the changes in the equation of time throughout the year. Take date horizontally, E vertically above and below the axis, plot the eight values given on page 25, and join them by a smooth wave-like curve. From the graph find (i) the value of E on March 1, (ii) the dates on which E = –6m. Suggested scales: 1 inch to a month and $\frac{2}{10}$ to a minute, or 2cm to a month and 3mm to a minute.

41 Convert the following times into angles: (i) 13h, (ii) 5h 36m, (iii) 11h 04m 28s.

42 Convert the following angles into time: (i) 120°, (ii) 60°12′, (iii) 81°37′45″.

43 If GMT is 12h what is the local mean time (i) in long. 80°E, (ii) at Harvard, USA, long. 71°34′W, (iii) Sydney, Australia, long. 151°09′E?

44 If it is 12h LMT at Harvard, long. +71°34′, what is the LMT (i) at Greenwich, (ii) at Palomar Observatory, long. +116°52′?

45 If it is 22h UT on 8 July what is the local time and date (i) at Greenwich, (ii) in zone +5, (iii) in zone −5?

46 If an aircraft leaves London at 11 pm British Summer Time and takes 7 hours to fly to New York (long. approx. 75°), at what LMT does it land?

47 Estimate the sidereal time at (i) 20h UT on 1 March, (ii) 04h 30m on 23 August.

48 Calculate the sidereal times at (i) 20h UT on 1 March, (ii) 04h 30m UT on 23 August, given that the sidereal times at 0h UT were 10h 33m and 22h 03m respectively.

49 On a day when the sidereal time at 0h UT was 6h 40m, what will it be at 22h 15m LMT (i) at Armagh Observatory, long. +6°39′, (ii) Radcliffe Observatory, long. −28°12′?

50 Given that the sidereal time at 0h UT is 04h 00m, (i) at what GMT will the sidereal time be 11h 45m, (ii) at what LMT in long. 100°E will the sidereal time be 11h 45m?

EXERCISES ON CHAPTER 3

Group 1: Practical

Improvised equipment has been assumed as usual, but institutions possessing a transit, a theodolite, or a sextant will be able to reach a higher degree of accuracy.

51 To find latitude with a shadow stick.

You require the same rod or post as used in Ex. 1. By the method of Ex. 27 calculate the mean time of solar noon, and while using the almanac extract the declination of the Sun. At exact noon measure the length of the shadow. Then the height of the stick divided by the length of the shadow is the tangent of the meridian altitude; if tangents are not available, draw to scale and measure the angle. Meridian alt. = (90° − lat.) + dec. Calculate the latitude.

52 To find meridian altitude with the clinometer.

Adequate protection for the eyes is important; dark glasses, smoked glass or old photo negatives are not really good enough. (Teachers with large classes are advised to make no exception to the rule 'never look at the Sun' and omit this exercise and the next.) Red and blue gelatine filters (Strand Electric 14 and 20) bound together with sticky tape will do for short periods of naked-eye work. If one edge is left unbound the combination will slip like an envelope over one of the sighting slots of the clinometer made for Ex. 3. Measure the altitude of the lower limb of the Sun, and add $\frac{1}{4}°$, its semi-diameter, to get the altitude of the centre. (If a sextant has been used, take a more suitable semi-diameter from the table on page 13.) Deduce the latitude from your result.

53 To time the transit of the Sun by direct observation.

Make a dark sleeve for the slits used for Ex. 2. Time the instant when first the preceding (west) limb is in line with the slits and then the following limb. Take the average and use it to calculate the equation of time as in Ex. 30.

54 To find latitude and longitude from a star.

Choose a bright star for which R.A. and dec. can be found from an almanac or elsewhere. Find the UT of its transit over the meridian slits, and immediately afterwards measure the meridian altitude with the clinometer. From the UT find the Greenwich sidereal time (page 31); the R.A. is the local ST; the difference is longitude in time; convert to angle. The latitude can be found in the same way as in Ex. 51. The use of a simple pair of slits is limited to objects at low altitude, say up to 30°. A possible alternative is a pair of plumb-lines (provided that there is no wind) as they can be much taller. Better still is to design for yourself and make a sighting device which will move up and down like a transit telescope.

55 To find the R.A. and dec. of a star.

This is the same as Ex. 54 except that you assume your own co-ordinates and determine those of the star.

56 To make an equatorial theodolite. (The name is Dr Tricker's. A more elaborate form is described in his book, and another will be found in Min. Ed. Pamphlet 38, *Science in Secondary Schools*, HMSO 1960.)

Make the object shown in Fig 38; details are left to you, for much of the pleasure of making things is in planning how to do it. For the purpose of illustrating parts of this chapter stiff cardboard will do, pins being used as pivots with corks behind, and circles hand-graduated in units of 10°. If some accuracy in measurement is intended wood must be used, and commercially made degree scales such as circular protractors. Angle ABC is equal to 90° – lat., so when the edge AB is placed along the meridian on a level surface the upper face will be in the plane of the equator. D is a disk pivoted at the centre and graduated in degrees anti-clockwise from 0 to 360°; it is read at the mark E. Mounted perpendicular to D and along the 0–180° diameter is another disk, F, marked in four 90° arcs, the line of the zeros being parallel with CB. G is a sighting alidade pivoted at the centre of F.

Place it horizontally on the meridian and using its two axes of rotation set the alidade so that the two points are in line with a star (or until the Sun casts a shadow of the southerly point on to the northerly one). Then scale F, read at either of the openings H, will be the declination; the sign should be obvious—it is south or negative in the diagram. The reading at E is the hour angle. But LHA in arc = H.A.♈ + SHA, and LHA in time =

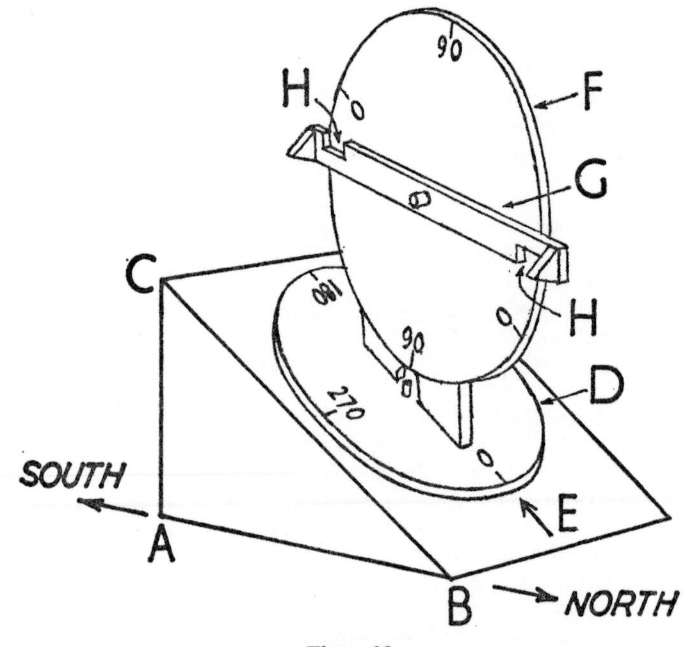

Figure 38

sidereal time – R.A. Hence SHA or R.A. can be found when you have calculated sidereal time (page 31).

An alternative exercise is to find out the LHA and dec. for some celestial object at some convenient observing time, set the scales, and see if at the appointed time you can locate the object. This is the principle of the equatorial telescope; here however there is a scale for sidereal time adjacent to scale D, and once this has been set the driving mechanism of the telescope maintains it for as long as desired.

57 To record the direction and time of sunset.

Each day for as long a period as possible observe the time and direction of the setting Sun, the point where the upper limb finally vanishes. Use a compass and correct for variation as in Ex. 29. Tabulate: Date; GMT; azimuth of sunset; amplitude, expressed in degrees N or S of due west.

Group 2: Recalling Previous Reading

58 Describe the path of the Sun across the sky in June and in December as seen from (i) the UK, (ii) Australia.

59 What is meant by the 'circumpolar stars'? Name some circumpolar constellations, and explain any changes which you would observe if you travelled from the UK to (i) the Arctic Circle, (ii) the Cape of Good Hope.

60 What and where are the Tropics of Cancer and Capricorn, and what is their astronomical significance?

Group 3: General

61 What are sidereal time and Right Ascension? Show clearly how they are related to each other.

62 Define meridian altitude, declination, zenith distance, polar distance. Point out any relationships between them.

63 What is meant by sidereal hour angle? Convert R.A. (i) 06h 30m, (ii) 18h 12m into SHA in degrees.

64 When the sidereal time is 10h find the local hour angle (in time units) of R.A. (i) 16h 32m, (ii) 22h 10m, (iii) 03h 50m.

65 When the sidereal time is 07h 41m 32s find the LHA of (i) Capella, R.A. 05h 14m 28s, (ii) Regulus, R.A. 10h 06m 47s.

66 Find the local hour angle of objects of GHA 70° and 270° in longitudes (i) 90°W, (ii) 60°E.

67 If the sidereal time at Greenwich is 16h 46m, find the LHA of a star in R.A. 05h 26m observed in long. 15°W.

68 If the GHA of a star is 53°17′28″, what will be the LHA at the same instant in (i) Rome, long. 12°27′06″E, and (ii) Toronto, long. 79°33′54″W?

69 The co-ordinates of Greenwich and Dublin are respectively $\theta = 0°$, $\phi = +51°28'38''$ and $\theta = +06°20'18''$, $\phi = +53°23'13''$. Find (i) the sidereal time at Greenwich when Capella, $\alpha = 05\text{h}\ 14\text{m}\ 28\text{s}$, $\delta = +45°58'10''$, is in upper transit at Dublin. Find also the meridian altitude of this star (ii) at Dublin, (iii) at Greenwich.

70 The solstices are the points on the ecliptic of greatest angular distance from the equator; the Sun passes them about June 21 and Dec 22. What is the declination on these dates, and what would be its meridian altitude when observed from (i) Manchester, lat. +53°29′, and (ii) Cape Town, lat. −33°56′?

71 What are the maximum and minimum meridian altitudes of the Moon, seen from latitude 52°N? At what times of year would you expect them to occur, and would they be regular annual phenomena?

72 What is azimuth? What are the altitude, azimuth and zenith distances of an object on the celestial equator (i) when it is rising, (ii) when it is crossing the meridian, the latitude being 50°N?

73 A star has R.A. 05h 30m, dec. +50°. What is the sidereal time of lower culmination, and what would then be its altitude for an observer in latitude 52°N?

74 Reconsider the last question for an observer in latitude 32°.

75 If the upper culmination of a star occurs at 07h 27m UT, what is the time of the next lower culmination?

76 In latitude 52°, if the altitude of a star at lower culmination is 10° what will it be at the upper? What do you notice about the average of the two altitudes when both are measured from the north?

77 Explain the meaning of the astronomical triangle PZX.

78 The Sun crosses the meridian at a certain place at 12h 05m UT and at an altitude of 45°, on a day when a sundial at Greenwich is 10m fast and the declination of the Sun is +5°. Find the latitude and longitude of the observer.

79 Draw a celestial hemisphere with the observer's horizon horizontal and in latitude 50°N. Insert the points N, S, E, W, the pole, the zenith, the equator, the approximate position of ♈ at 21h LMT on September 23, and a star at R.A. 18h, dec. +30.

80 Draw a celestial hemisphere as in the last exercise. Insert the equator and the daily paths of the Sun on June 21 and Dec. 22 What do you think would be the approximate azimuth of the rising and setting points on these two dates? What would be the corresponding amplitudes?

81 Draw a celestial sphere showing the poles, equator and ecliptic. Label the points ♈ and ♎. On the equator write X at about R.A. 6h, Y at about SHA 120°, and on the ecliptic Z at long. 90°.

82 Repeat Ex. 79 for an observer in lat. 30°S.

83 The first two 'soft landings' on the Moon were made by Luna 9 and Surveyor 1. Their camera lenses were respectively 60cm and 160cm above the surface. Assuming that the surface was uniform (which it was not) and that the radius of the Moon is 1740km, how much further could Surveyor see?

84 Given that the radius of the Earth is 6370km, what is the distance of the Moon on a day when the horizontal parallax is 54′?

85 Draw a celestial sphere, showing the equator and the poles, and

the ecliptic and its poles. What are (i) the R.A. and dec. of the poles of the ecliptic, (ii) the longitude and latitude of the celestial poles? Draw any great circles which you have used in thinking this out.

EXERCISES ON CHAPTER 4

In certain cases junior and senior have been differentiated

Group 1 : Practical

86 Observe the planets regularly for as long a period as possible, locate them as well as you can among the stars, and record their positions on a star map. For Venus, Mars, and Jupiter when near opposition, intervals should not be longer than a week. (If an equatorial theodolite is available you might like to measure their positions and plot them on squared paper.)

86A Instead of estimating the position of a planet by eye, measure its distance from two neighbouring stars by means of a cross staff, Fig 39, the instrument used for such work by early

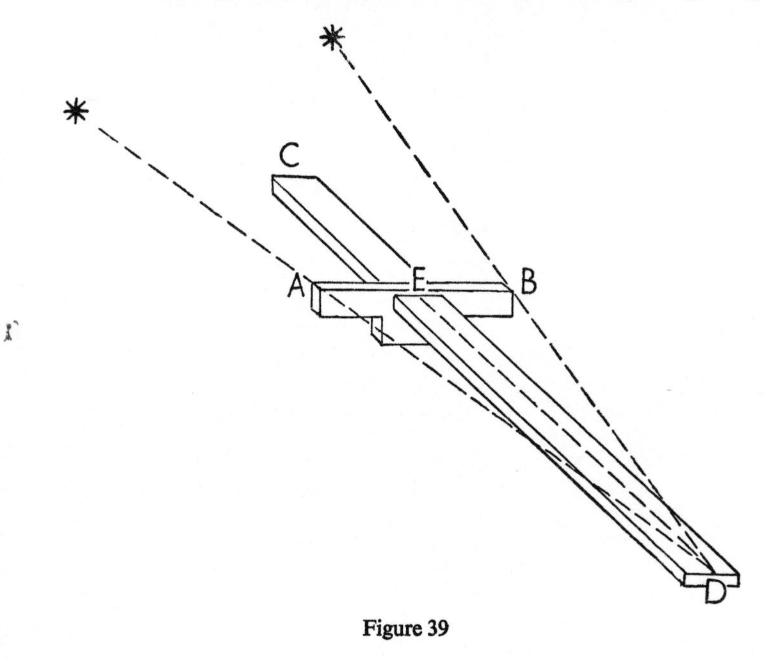

Figure 39

navigators. AB slides along CD and D is held close to the eye, so the further AB the greater the angle subtended by it. A little light will be needed to see the cross piece against the dark sky, say by working a few yards from a lighted window. A laboratory rule can be used for CD.

To determine the angle: (a) AB/DE = angle in radians (provided that the angle is fairly small, see p. 4); convert to degrees.

Example: If AB = 10cm and DE = 60cm,

$$\text{angle ADB} = \frac{10}{60} = 0{\cdot}167 \text{ radian} = \frac{0{\cdot}167 \times 180}{\pi} = 9^\circ{\cdot}55.$$

(b) If you are familiar with trigonometrical tables, EB/ED = tan the half angle BDE, and the example becomes

$$\tan \text{BDE} = \frac{5}{60} = 0{\cdot}0835 \quad \text{BDE} = 4^\circ 46'$$

$$\text{ADB} = 2 \times 4^\circ 46' = 9^\circ 32' = 9^\circ{\cdot}53$$

(c) Use a plain rod for CD and calibrate in degrees, using the following table. To find DE multiply AB by the number in column 2. By using two different lengths of the cross piece, as shown in the diagram, you can make two scales such as 1–7° on one side and 7–30° on the other.

θ	$\frac{1}{2}\cot\frac{1}{2}\theta$	θ	$\frac{1}{2}\cot\frac{1}{2}\theta$
1°	57·3	9°	6·3
2	28·7	10	5·7
3	19·1	12	4·7
4	14·3	15	3·8
5	11·4	18	3·1
6	9·5	20	2·8
7	8·2	25	2·2
8	7·1	30	1·8

87J By the method of Ex. 2 find the time of transit of one of the planets, and hence fix its position on your orbit diagram (Ex. 100). Repeat at intervals, and so follow the motion of the planet along its orbit.

88S *Either* by the method of Ex. 53–54 find the sidereal time of transit of a planet and of the Sun on the same day; these will be their R.A., *or* measure their R.A. with the equatorial theodolite. Find their longitudes from the table on page 63 and insert them on your orbit chart, Ex. 104 or 105. (Remember that heliocentric longitude of the Earth = geocentric longitude of the Sun ± 180°.)

Group 2: Recalling Previous Reading

89 Describe several ways of recognising a planet in the sky.

90 Explain the meaning of this imaginary extract from the 'Stars of the Month' in a newspaper: 'Venus is an evening star and reaches elongation on the 7th. Mars, in superior conjunction on the 18th, will not be observable this month. Jupiter is still a morning star until opposition on the 29th.'

91 Why are there phases of Venus but not of Jupiter?

Group 3: General

92 The following table gives monthly positions of the Sun for a year and for Venus for two years (such a table is called an *ephemeris*). Plot the points on squared paper; number the points as you go along, in pairs from 1–24; join with smooth lines, preferably in different colours; on the Venus track write 'morning star' and 'evening star' as appropriate. (As a shorter exercise plot 1970 July to 1971 March only.) Suggested scales on foolscap size paper and arranged like the specimen map on page 84: $\frac{1}{2}$ inch to an hour of R.A., $\frac{1}{10}$ inch to a deg. of dec.; 1cm to an hour, 2mm to a deg. In each case the dec. scale is three times that of R.A. On this scale the graph for the Sun is nearly the same every year and need not be drawn twice; give the same points both sets of numbers, 1–12 and 13–24.

1st	SUN			VENUS 1970			VENUS 1971		
of	R.A.		Dec.	R.A.		Dec.	R.A.		Dec.
Jan	18h	44m	−23°	18h	19m	−23½°	15h	33m	−15½°
Feb	20	57	−17	21	05	−18	17	39	−20
Mar	22	46	−8	23	20	−6	19	53	−19½
Apr	00	40	+4½	01	41	+10	22	20	−11
May	02	31	+15	04	06	+21½	00	35	+2
June	04	34	+22	06	50	+25	02	57	+15½
July	06	38	+23	09	19	+17½	05	29	+23
Aug	08	43	+18	11	31	+3½	08	13	+21
Sep	10	39	+8½	13	26	−11½	10	45	+9½
Oct	12	27	−3	14	59	−22½	13	02	−5½
Nov	14	23	−14	15	14	−24	15	30	−19
Dec	16	26	−21½	14	30	−14	18	10	−24½

93 Here is the ephemeris for Jupiter for one year (1970).

Jan	14h	02m	−11	May	13h	53m	−10	Sep	14h	03m	−11½
Feb	14	14	−12	June	13	41	−9	Oct	14	24	−13½
Mar	14	16	−12	July	13	38	−9	Nov	14	50	−15½
Apr	14	07	−11½	Aug	13	46	−10	Dec	15	16	−17

Scales: 3 inches to an hour of R.A.; $\frac{2}{10}$ inch to a deg. of dec.; 6cm to an hour, 4mm to a deg. These are equal scales. Plot the points, join by a smooth curve, write the date against each point, and estimate the date of opposition.

94 The synodic period of Venus is 584 days. Calculate the sidereal period.

95 Jupiter was in opposition on 1969 March 21. Calculate the date of the next one. Sidereal period 11·9 years (4 333 days).

96 Use these satellites of Uranus to illustrate Kepler's third law: 131 000km, 1·41d; 192 000km, 2·52d; 268 000km, 4·14d; 586 000km, 13·5d.

97 The distances of the satellites of Mars are: Phobos, 9 370km, and Deimos, 23 600km. If the period of Phobos is 7·65 hours, what is that of Deimos?

98 A 'hovering' artificial satellite remains over the same region of the world because it has a period of 1 day. Given that the distance of the Moon is 384 000km, its period 27·3 days, and the radius of the Earth 6 370km, find the altitude of the satellite above the surface.

99 Write the numbers 0, 3, 6, 12 etc. doubling each time until you have eight altogether. To each add 4, making 4, 7,.... Underneath write the distances of the planets, omitting Eros, from page 61. Study the list; can you see a better way of writing it? (This is known as *Bode's Law*; it is doubtful whether it has any theoretical significance.)

100J Draw the orbits of Mercury, Venus, Earth and Mars as explained on page 61. Show the planets in the positions they would occupy on July 1 if their transit times were Mercury 11h 35m, Venus 14h 45m, Mars 12h 47m. Mercury was after and Venus before elongation.

101J Transits on June 1 were Jupiter 21h 00m, Saturn 10h 18m, Uranus 19h 40m, Neptune 23h 00m. Draw their orbits and that of the Earth as explained on page 61 and find the positions of the planets.

102J Draw the orbits of the Earth and of Venus. Venus crossed the meridian at 09h 12m on Dec 16; place it in its orbit, measure its geocentric longitude, and state in what constellation it would be. Inferior conjunction was on Nov 10.

103J Draw the orbits of the Earth and Jupiter. Jupiter crossed the meridian at 17h 08m on Aug 1; place it in its orbit and measure its distance from the Earth in astronomical units.

104S Using the instructions on page 61 and the data on page 61, draw the orbits of the four nearer planets (excluding Eros) and place them in their orbits for 1970 July 1. Do not place Mercury, as the daily rate varies too much.

105S The same as 104, but for the Earth and the four great planets on June 1.

106S The maximum elongation of Mercury varies considerably. Using drawing 104, by inspection and a few trial measurements find the greatest and least values.

107S Using drawing 105 find the longitude and latitude of Saturn on 1970 Mar 5, and with the help of the map on page 84 convert to R.A. and dec.

108S Using drawing 104 find the longitude and latitude of Venus on 1970 June 13, and deduce the constellation in which it would lie.

109 If the necessary data is available and by junior or senior methods as appropriate, find the positions and constellations of the naked-eye planets for tomorrow. Consider which ought to be visible in the evening sky and in what direction. When tomorrow comes, go out and look for them.

110S Add the orbit of Eros to drawing 104. If its nearest approach to the orbit of the Earth is 21×10^6km, find a value for the astronomical unit.

111 Adonis, like Eros in the last question, is an asteroid, but with such a large eccentricity that it must be drawn as an ellipse. Add as much as you can of it *either* to drawing 100J (the perihelion of the orbit is in long. 32°, $a = 2{\cdot}0$, $e = 0{\cdot}78$); *or* to drawing 104S ($\varpi = 32°$, ☊ $= 353°$, $i = 1°$, $e = 0{\cdot}78$, $a = 2{\cdot}0$, $q = 0{\cdot}44$, $P = 2{\cdot}5$y). Choose for yourself the method of drawing the ellipse.

112J Draw the orbit of Encke's comet from instructions on page 67. ☊ $= 335°$, $\omega = 185°$, $a = 2{\cdot}21$, $e = 0{\cdot}85$. Draw the orbit of the Earth (a circle will do) on the same paper, and mark the position of the planet when the comet was at perihelion on 1971 Jan 19. Ellipse method 2 suggested.

113S Draw the orbits of Encke's comet, the Earth and Mars, three-dimensionally, on two cards from the data in 112 and $i = 12°$. Planet orbits can be drawn as concentric circles for this exercise and method 4 (p. 103) is suggested for the ellipse. The distance of the comet from the Sun on 1970 Nov 30 was 1·13A.U. Mark the positions of the Earth and the comet on this date. The period of this comet is 3·3y.

114J Draw the orbit of Halley's comet and on the same paper those of the Earth and Jupiter. Draw it as a parabola by ellipse method 1 to a distance of about 6A.U. from the Sun. ☊ $= 58°$, $\omega = 112°$ measured clockwise (see page 69), $q = 0{\cdot}587$, $e = 1$.

115S The same as 114 but three-dimensionally; $i = 162°$. Indicate the directions of motion of the comet and the planets. The eccentricity is really 0·97; the period of this comet is 76y.

116S Working on drawing 104S find the approximate date and

direction (radiant) of a meteor shower having these elements: $\Omega = 138°$, $\omega = 155°$, $q = 0{\cdot}96$, $e = 0{\cdot}96$, $i = 116°$.

117 State Newton's law of gravitation and Kepler's third law, and derive a relationship between them.

118 What is the force between masses of 8kg and 2g when placed 80mm apart? ($G = 6{\cdot}7 \times 10^{-11}$ S.I.). In what unit is your answer? Suggest why the experimental determination of G calls for a high degree of refinement.

119 Satellite II of Jupiter is at a distance of 0·0044A.U. and its period is 3·6 days. The mass of the satellite is 1/40 000 of that of its primary. Find a value for the mass of Jupiter; in what units is it?

120 By reference to the table on page 61 what is the distance of Saturn from the Earth at the time of opposition (i) in A.U., (ii) in km? If the angular diameter of the rings is then 46″ what is (iii) their diameter in km? (Hint on page 5.)

121 An object is projected tangentially (at right angles to the direction of the Earth) from a point X, situated 10 000km from the centre of the Earth, with a velocity of 7·0km s^{-1}. Show whether the orbit will be (a) circular, (b) an ellipse within the circle, (c) an ellipse outside the circle. If the answer is (b) or (c) state the position of perigee.

122 Calculate the velocity of escape from the surface of the Moon, assuming that its mass is 1/80 of that of the Earth (p. 74). (Radius of Moon 1 740km; $G = 6{\cdot}67 \times 10^{-11}$ S.I. units.)

EXERCISES ON CHAPTER 5

Group 1: Practical

123 If an equatorial theodolite (Ex. 56) is available use it to fix the positions of all the stars in a constellation and draw a chart from your results. This exercise is discussed in detail by Tricker.

124 The following stars are very near to integral (whole number) magnitudes. If a star atlas is available locate them in the sky as opportunity allows; they are fairly well spread.

Mag. 4	ι Leonis, η Cygni
Mag. 3	ϵ Persei, γ Bootis, δ Cygni
Mag. 2	Polaris, Hamal (α Arietis), ζ Orionis
Mag. 1	Spica, Pollux, Fomalhaut
Mag. 0	Vega, Capella, Arcturus, Rigel

The nearest to −1 is the southern star Canopus, −0·71; Sirius is −1·47.

125 Fig 40 shows two groups of stars with the magnitudes of some of them written alongside. Find these groups in the sky and estimate the magnitudes of stars not so labelled.

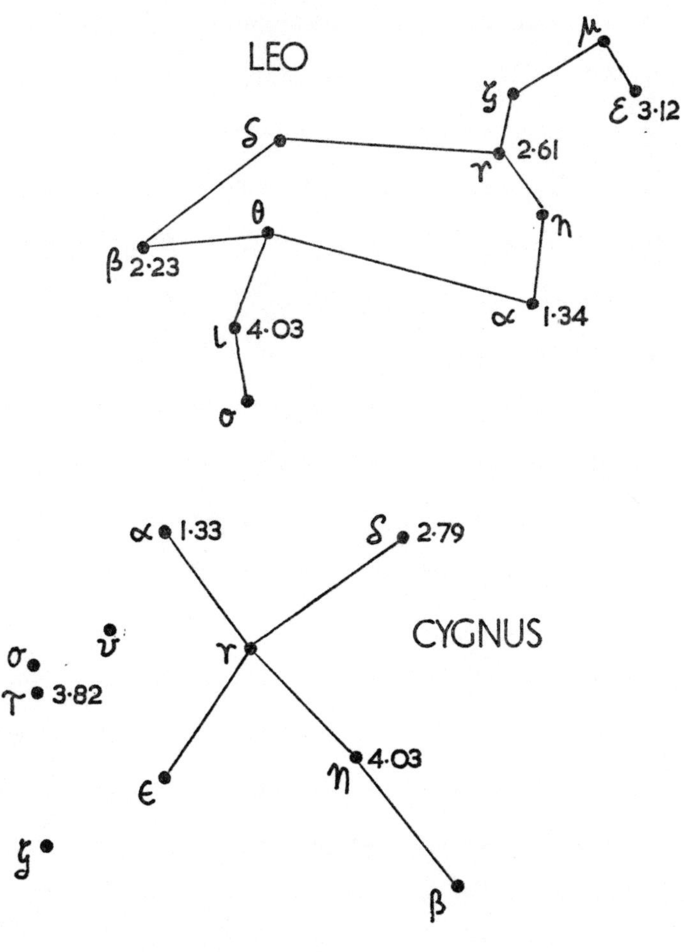

Figure 40

126 Fig 41 is the constellation of Lyra, β of which is variable; the magnitudes of the other stars are quoted. As often as possible for four or five weeks try to estimate the magnitude of β; for instance, if you consider it to be brighter than δ but less bright

than ζ, call it 4·4. Regular variable star observers can work to about a fifth of a brightness interval. Plot your results on graph paper, with date horizontal and magnitude vertical. Can you deduce a period? When you have finished (but *not* before) look for the light curve of this star in other books.

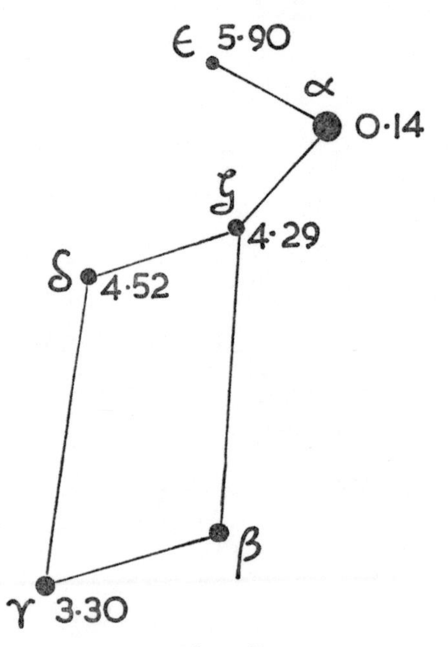

Figure 41

Group 2: General

Questions marked L involve the use of logarithms.

127–8 Draw charts, on the rectangular system, of the stars listed below. Name any stars or constellations which you recognise.

127 R.A.	127 Dec. all +	127 Mag.	128 R.A.	128 Dec.	128 Mag.
09h 43m	24°·0	3	04h 33m	+16°·4	1
50	26·2	4	48	+6·8	3
10 04	17·0	4	49	+6·5	4
06	12·2	1	52	+2·3	4
14	23·6	4	05 05	−5·1	3

127

R.A.		Dec. all +	Mag.
	17	20·1	2
	30	9·5	4
11	11	20·8	3
	12	15·6	3
	19	6·3	4
	21	10·8	4
	47	14·8	2

128

R.A.		Dec.	Mag.
	12	−8·2	1
	15	−6·9	4
	22	−2·3	3
	22	+6·3	2
	30	0	2
	32	+9·9	4
	34	−1·2	2
	36	−2·7	4
	38	−2·0	2
	45	−9·7	2
	52	+7·4	1
06	20	−17·9	2
	43	−16·6	−1

129 Draw a chart of the principal south circumpolar stars, using the polar system. Link up stars belonging to the same constellation and write in names and Greek letters. (α Eridani, Achernar; α Carinae, Canopus; α Crucis, Acrux; α Centauri, Rigil Kent).

Star		R.A.		Dec. all −	Mag.
Hydrus	β	00h	23m	77°·5	3
	α	01	57	61·8	3
	γ	03	48	74·4	3
Eridanus	α	01	46	57·5	2
Reticulum	α	04	14	62·6	3
Dorado	α		33	55·1	3
Pictor	α	06	48	61·9	3
Carina	α		22	52·7	−1
	ε	08	20	59·3	2
	β	09	13	69·5	2
	ι		15	59·0	2
	q	10	15	61·1	3
	θ		41	64·1	3
Vela	δ	08	43	54·5	2
	κ	09	21	54·8	3
Crux	δ	12	12	58·5	3
	α		23	62·8	1
	γ	12	28	56·8	2
	β		45	59·4	2
Musca	α		34	68·8	3
	β		43	67·8	3

Star		R.A.		Dec. all –	Mag.
Centaurus	λ	11	34	62·7	3
	ϵ	13	37	53·2	3
	β	14	00	60·1	1
	α		36	60·6	0
Circinus	α		38	64·8	3
Triangulum A.	γ	15	14	68·5	3
	β		50	63·3	3
	α	16	43	68·9	2
Ara	ζ		54	55·9	3
	β	17	21	55·5	3
	α		28	50·0	3
Pavo	α	20	21	56·9	2
Toucan	α	22	15	60·5	3

130 The magnitudes of the naked-eye stars run from –1 to +5. Estimate the ratio of brightness which this means.

131L Calculate the brightness ratio between (i) the twins, Castor 1·56, Pollux 1·15; (ii) the top two, Sirius –1·47, Canopus –0·71.

132L The star Algol (β Persei) varies in magnitude from 2·2 to 3·5. What brightness ratio is this?

133L Castor is a double star of mag. 1·56. If the brighter component has a magnitude of 2·0 what is that of the fainter?

134A If one star is 300 times as bright as another, estimate the difference in magnitude.

134BL As above, with 'calculate' instead of 'estimate'.

135 (i) What is the distance in parsecs of a star having a parallax of $0''\cdot02$? Aldebaran has a parallax of $0''\cdot048$; find its distance in (ii) parsecs, (iii) astronomical units.

136 (i) What is the parallax of a star 32·6 light years away? (ii) If the distance of Regulus is 85 light years what is its parallax?

137L Calculate the absolute magnitude of (i) Antares, apparent mag. 0·92, distance 122pc; (ii) Canopus, apparent mag. –0·71, distance 92pc.

138L The apparent stellar magnitude of the Sun is –26·8. Taking its distance to be 1/206 000pc calculate its absolute magnitude. ($\log 1/206\,000 = \bar{6}\cdot686 = -5\cdot314$.)

139L Using results from the last two questions find the ratio by which Canopus is brighter than the Sun.

140L The absolute magnitude of Acrux, deduced from spectroscopic observation, is –3·8 and the apparent magnitude is 0·8. Find its distance (i) in parsecs, (ii) in light years.

141 Sirius is a binary star with a period of 50 years, true separation $7''\cdot6$ and parallax $0''\cdot38$. Calculate the mass of the system.

142 A binary star has a period of 34 years and a true separation of 1″·35. Assuming the mass to be 2·0 sun-masses, calculate (i) the parallax, (ii) the distance in parsecs.

143 What is the frequency of (a) the 21cm line used by radio astronomers, (b) the sodium line at 5 890Å?

144 If a spectral line known to be 5 000Å is observed at 5 020Å, what is the velocity of the source?

145 The K line of calcium has a wavelength of 3 934Å. What would be the observed wavelength in the spectrum of a galaxy receding with a velocity of 15 000km s^{-1}?

Answers

p. 113	7	0·0033. The 'table book value' is $\frac{1}{298}$
	8	(i) $\phi = +55°56'$, $\theta = +3°11'$
		(ii) $= +45°30'$, $+73°35'$
		(iii) $= +34°07'$, $+118°18'$
		(iv) $= -33°50'$, $-151°10'$
		(v) $= +55°45'$, $-37°34'$
		(vi) $= -26°11'$, $-28°05'$
	9	(i) 17°12′ } fourth figure doubtful
		(ii) 63°02′ } fourth figure doubtful
		(iii) 0·524
		(iv) 0·828
	10	17′·2
	16	☊ = 328°; $e = 0·055$
	17	(i) July 1
		(ii) Jan 1
		(iii) Nov 9
		(iv) Nov 23
	18	Annular eclipse
	22	1 368 000km
	23	(i) 377 800km
		(ii) 38·5km
	24	(i) 9 170km
		(ii) 3 640km per hour
		(iii) 0·6h or 36m
	26	(i) solar–lunar–solar
		(ii) one solar only
p. 119	38	11h 52m mean time, 12h solar time
	39	(i) 07h, 17h
		(ii) 17h 06m, 17h 06m
	40	Almanac values for 1970:
		(i) −12m 27s
		(ii) Jan 6, Mar 25, July 17, Aug 3 (the maximum being −6m 26s on July 26)
	41	(i) 195°
		(ii) 84°
		(iii) 166°07′

42 (i) 8h
(ii) 4h 00m 48s
(iii) 5h 26m 31s
43 (i) 17h 20m
(ii) 7h 13m 44s
(iii) 22h 04m 36s
44 (i) 16h 46m 16s
(ii) 8h 58m 48s
45 (i) 22h July 8
(ii) 17h July 8
(iii) 03h July 9
46 00h LMT (midnight)
47 (i) 6h 24m
(ii) 2h 30m
48 (i) 6h 36m 20s
(ii) 2h 33m 40s
49 (i) 4h 58m 40s
(ii) 4h 58m 20s
50 (i) 7h 43m 50s
(ii) 7h 44m 50s

(48–50: The additional correction at the bottom of page 30 has not been used.)

p. 123 63 (i) 262°30′
(ii) 87°
64 (i) 17h 28m
(ii) 11h 50m
(iii) 6h 10m
65 (i) 2h 27m 04s
(ii) 21h 34m 45s
66 (i) 340°, 180°
(ii) 130°, 330°
67 10h 20m or 155°
68 (i) 65°44′34″
(ii) 333°43′34″
69 (i) 5h 39m 49s
(ii) 82°34′57″
(iii) 84°29′32″
70 (i) June 60°01′, Dec 13°01′
(ii) June 32°34′, Dec 79°34′
(Taking $\epsilon = 23°26'35''$ the answers would be 59°57′35″, 13°04′25″, 32°37′25″, 79°30′35″)
71 $66\frac{1}{2}°$ in northern winter, $9\frac{1}{2}°$ in summer
No, on account of the motion of the Moon's nodes
72 (i) alt. = 0, az. = 90°, zd = 90°
(ii) 40°, 180°, 50°
73 17h 30m, 12°
74 No lower culmination because the zenith distance is greater than the altitude of the pole
75 19h 25m UT
76 86° from the south horizon. This would be 94° from the north, and the average of 94 and 10 is 52, the latitude of the observer
78 Lat. 50°N, long. 3°45′W

	80	Suppose you guessed the June rising and setting to be 30°N of E and W; then the azimuths would be 060° and 300° and amplitudes both 30°N. The December values would be 120°, 240°, 30°S. The calculated amplitude is 38°
	83	Horizon distances 2·36 and 1·45km, giving a difference of 0·91km
	84	405 600km
	85	(i) R.A. 18h, dec. $66\frac{1}{2}$°N; R.A. 6h, dec. $66\frac{1}{2}$°S (ii) Long. 90°, lat. $66\frac{1}{2}$°N; long. 270°, lat. $66\frac{1}{2}$°S
p. 128	94	225 days
	95	Simple calculation in whole days gives Apr 23; it was actually 1970 Apr 21
	96	The ratios d^3/T^2 come to 1·13, 1·11, 1·12, $1{\cdot}10 \times 10^{15}$
	97	30·6 hours
	98	42 350km from the centre; altitude 35 980km

99

0	3	6	12	24	48	96	192
4	7	10	16	28	52	100	196
0·39	0·72	1·0	1·5		5·2	9·5	19·2

The space at 28 represents the asteroid belt

100 to 116 Geometrical constructions as described will not give exact results, but to check whether they are reasonable, almanac or calculated values are given below. Some answers are sketched in Fig 42

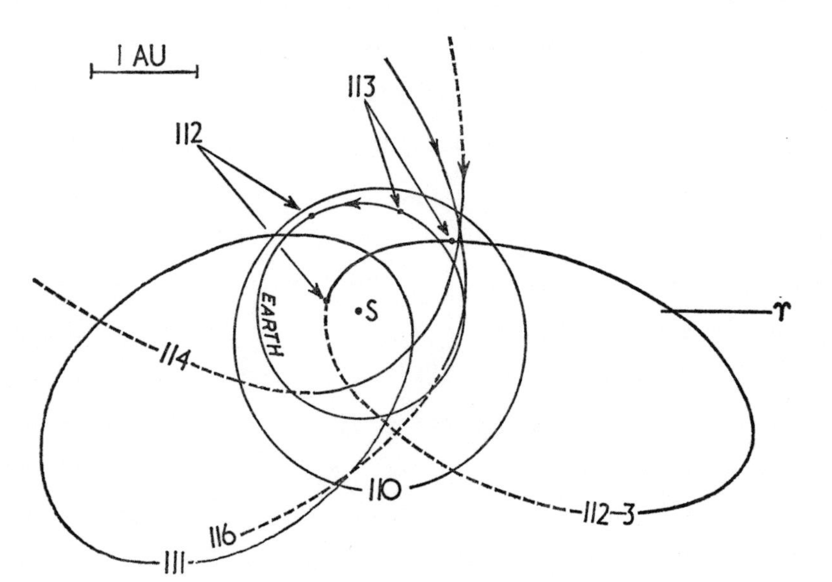

Figure 42

	100	Heliocentric longitudes: Mercury 067°, Venus 197°, Earth 279°, Mars 115°
	101	Heliocentric longitudes: Jupiter 214°, Saturn 043°, Uranus 187°, Neptune 239°
	102	Heliocentric long. 105°, geocentric long. 225°, in Libra
	103	Heliocentric long. 219°, distance 5·51A.U.
	104	Same as Ex. 100
	105	Same as Ex. 101
	106	About 18° to 28°
	107	Geocentric long. 036°, lat. −2°; R.A. 02h 15m, dec. +11°
	108	Geocentric long. 116°, lat. +2°; in Gemini at R.A. 07h 52m, dec. +23°
	110	1A.U. = 149 000km
	116	The scattered Perseid stream is encountered from about July 25, R.A. 02h 40m, dec. +56°, to about Aug 17, R.A. 03h 10m, dec. +58°; maximum Aug 12
	118	$1{\cdot}68 \times 10^{-10}$ newton. The force between laboratory masses such as these is very small and difficult to measure.
	119	0·000 876 sun-mass or 292 earth-masses
	120	(i) 8·54A.U. (ii) $1{\cdot}28 \times 10^{9}$km (iii) 285 600km
	121	Case (c); perigee at X
	122	2·39km s^{-1}
p. 131	125	Leo: μ 4·1, η 3·58, ζ 3·65, δ 2·58, θ 3·41, σ 4·13 Cygnus: β 3·1, γ 2·32, ϵ 2·64, ν 4·04, σ 4·28, ζ 3·40
	126	Mag. 3·4 to 4·3; period 12·9 days
	130	250 approx.
	131	(i) 1·46 (ii) 2·01
	132	3·31
	133	2·75
	134	6·19
	135	(i) 50pc (ii) 20·8pc (iii) $4{\cdot}3 \times 10^{6}$A.U.
	136	(i) 0″·1 (ii) 0″·038
	137	(i) −4·51 (ii) −5·53
	138	+4·77
	139	26 000
	140	(i) 83·2pc (ii) 272l.y.
	141	3·2 sun-masses
	142	(i) 0″·102 (ii) 9·8pc
	143	(a) $1{\cdot}43 \times 10^{9}$Hz or 1 430 megacycles per sec (b) $5{\cdot}1 \times 10^{16}$Hz
	144	12×10^{5}m s^{-1} or 1 200km s^{-1}
	145	4 130Å

Index